Traces of Common Xylophagous Insects in Wood

Magali Toriti • Aline Durand • Fabien Fohrer

Traces of Common Xylophagous Insects in Wood

Atlas of Identification - Western Europe

Magali Toriti
Centre de Recherche en Archéologie, Archéosciences,
Histoire (CReAAH UMR CNRS 6566)
Le Mans University
Le Mans, France

Aline Durand
Centre de Recherche en Archéologie, Archéosciences,
Histoire (CReAAH UMR CNRS 6566)
Le Mans University
Le Mans, France

Fabien Fohrer
Centre Interdisciplinaire de Conservation et de Restauration du
Patrimoine (CICRP)
Marseille, France

ISBN 978-3-030-66393-3   ISBN 978-3-030-66391-9   (eBook)
https://doi.org/10.1007/978-3-030-66391-9

This Springer imprint is published by the registered company Springer Nature Switzerland AG
The registered company address is: Gewerbestrasse 11, 6330 Cham, Switzerland

Foreword

A Professional Passion Turned into a Much-Needed Book

I was a student in wood science when I stumbled on wood-boring insects. I loved thinking, talking and studying about wood-boring insects. I even thought of a possible career in entomology. But the prospect of drilling deeper into a single academic field felt confining. So, I ended up building my academic career around the anatomy and ecology of plant stems. I do teach about these topics too. It was while teaching at an International Course on Wood Anatomy and Tree-Ring Ecology held in Switzerland that I met Magali Toriti. At that time she was a PhD student in archaeology supervised by Aline Durand. Both Magali and Aline were curious about some small circular holes in charcoal samples. But both didn't know if it was worth exploring further into them. Thus we compared the holes in the charcoal with the material available in the teaching slide collection. And we found similarities to traces from wood-boring insects. To further meet Magali scientific curiosity the expertise in entomology of Fabien Fohrer came into play. Some years later Fabien also attended the same course on Wood Anatomy. Once I got acquainted with both Magali and Fabien, I felt so fortunate to share their motivation in understanding wood-boring insects with my old passion for the topic. Together we spent much time collecting insects, looking at wood samples, cutting thin anatomical sections and learning from the microscope.

The result of these efforts is summarised in the *Traces of Common Xylophagous Insects in Wood. Atlas of Identification—Western Europe* which gathers the knowledge they collected over the past eight years of collaboration. The book, after presenting some distinctive biological features of various types of wood pests and introducing wood structure, details the life cycle of a wood-boring beetle and provides a guide to using the identification keys. Two identification keys are introduced. The first focuses on the galleries and faecal pellets of wood-boring insects and is useful for applications when the damaged piece of wood is still visible, such as a museum object. The second key describes the sole faecal pellets of wood-boring insects. It helps when the wood is completely decayed or not readily available, such as in archaeological wood. Later (chapter 3), the bulk of the book describes the most common xylophagous insects. They are described by family within the Coleoptera, Curculionidae, Lyctidae, Ptinidae, and Hymenoptera and Isoptera.

This book represents what I was desperately looking for at the beginning of my career: a guide for the beginner into the world of wood-boring insects. But the book is also a valid tool for the expert of the field looking for a comprehensive description of the problem of wood pests.

Department of Geography
University of Cambridge,
Cambridge, UK

Alan Crivellaro

Acknowledgements

We thank Alan Crivellaro and Lisa Shindo for their wood samples, Christian Cocquempot for checking the Cerambycids table, Hervé Bouyon for his photos of some Cerambycids, Loïc Joanny and Francis Gouttefangeas (CMEBA-Scanmat) and Alain Tonetto and Samuel Saller (Pratim) for their precious help in SEM photography. We would also like to thank Annie Buchwalter for english translation of the book.

Fundings

- Le Mans university (France) and the Center of Archaeology, ArchaeoSciences and History UMR CNRS 6566 (France)

- Pays de la Loire Region (France)
- The Earth Sciences and Astronomy Observatory of Rennes (OSUR Rennes I University)
- The Interdisciplinary Center for Heritage Conservation and Restoration (CICRP Marseille)

Our funders are:

- Le Mans University
- CICRP Interdisciplinary Center for Conservation and Restoration of Heritage (Marseille)

- Région Pays de la Loire (PhD Grant)

- OSUR Observatory of Sciences of the Universe of Rennes (MEB Photographs)

- Technological and Imaging Analytical Research Platform of University of Aix-Marseille (MEB photographs)

Photo Credits

Unless otherwise stipulated, all photographs, illustrations and diagrams are by the authors: insects imagi by Fabien Fohrer, galleries and frass by Magali Toriti. Several pictures of insect were given by Hervé Bouyon. SEM photographs were taken at the CMEBA-Scanmat laboratory by Loïc Joanny and Francis Gouttefangeas at the University of Rennes 1 and at the Plateforme de Recherche Analytique Technologique et Imagerie (Pratim) by Alain Tonetto (Ingetude) and Samuel Saller (technician) at Aix-Marseille University (AMU) Saint-Charles.

Contents

Introduction

Wood-boring insects are natural pests of trees, stored wood or timber. In forests, they participate in the partial or total degradation of dead wood. Depending on the species, they can be major pests of forests, of a given woody species population, or even destroy museums and wooden heritage collections. As for wood, it is a key element of the economy and the environment, both past and present.

In Western Europe, fundamental works on wood anatomy and keys for determining the comparative anatomy of insects only emerged in the last 80 years. They described the great variability of the discriminating characteristics of wood (Huber and Rouschal 1954; Greguss 1955; Brazier and Franklin 1961; Jacquiot et al. 1973; Couvert 1977; Schweingruber 1990; Vernet et al. 2001; Ruffinatto and Crivellaro 2019) and in insects and other arthropods (Lepesme 1944; Bremond and Lessertisseur 1973; Hoffmann 1986; the XIXth c. Die Kafer Europas collection, or the fauna collections "Faune de France", "Fauna Iberica", "Fauna Italiana", *per* country (e.g. Berger 2012)).

The starting point of this book was the analysis of charred wood from an alpine archaeological Gallo-Roman site. The rural settlement, well preserved according to archaeological criteria, was burnt down in the third century AD. As a result, various wooden construction elements were preserved and with them numerous traces of wood-boring insects (galleries and frass). In the absence of elements from the adult insect(s), it was not possible to identify the agent(s) of such an infestation, which could affect *pro parte* more than a third of the structure. Within the community of bioarchaeologists, many have noticed these traces, noted as "holes" on both charred and waterlogged wood, but few have attempted to identify the insect(s) that bored them. Most of the time, the more or less hazardous conservation of archaeological woody heritage makes the task even more difficult. For example, the ship The Mary Rose, discovered in the 1960s, was infested by *Nacerdes melanura* from the 1990s, when she was stored in a wet warehouse (Pitman et al. 1993). It was only by comparing these archaeological traces with those of current insects, whose exact species are known, that identification was made possible.

For a decade now, various tests have been carried out to identify the traces left by larvae on/in the wood. These tests concern only a few large families of woodborers. Sometimes it is necessary to evaluate and quantify the damage caused by an insect to a structure before insecticide treatment, in particular by observing the number of flight holes and trying to identify them due to the absence of an adult individual. Within the museum environment, one of the rare cures used until the 1950s consisted of regular checks of the collections in order to limit damage. Subsequently, notably with the commercialisation of organo-chlorinated insecticides among others, and especially the awareness of their toxicity for humans, interest in and knowledge of wood-boring insects and lignivorous fungi increased, with the aim of better preserving the woody heritage while eliminating pest threat. Thus, beyond museums, these methods are opening up and adapting to different media (ancient carpentry *in situ*, organ pieces in a church, current furniture and architecture, underwater wrecks ...). A few rare works (Gambetta 2010; Blanchette 1991; CTBA 1996; Bobadilla et al. 2015) or mentions (Lepesme 1944; Español 1992) underline the importance or specificity of the traces left by the larvae of wood-boring insects in the wood, i.e., galleries and larval dejections.

This book is based on the knowledge acquired by crossing over two disciplines, i.e., archaeology and heritage conservation-restoration. This crossover led to the creation of an original identification key, applicable to these two scientific fields. In addition to these two sectors, the key is also applicable to other fields: forest protection, wood industry and engineering, building and public works, or in the judicial field, for example in a dispute due to an infestation after or before property purchase. In the field of heritage conservation, insects damage works and buildings, making them a source of concern. In view of the diversity of species of wood-boring insects, their identification makes it possible to choose the best way to protect the object, structure, building, etc. Similarly, in forests, it is essential to know the causal agent of the decline of a woody population in order to better control it and face up with the situation.

There are currently more than 600 species of xylophagous insects in Western Europe. Each of these species preferentially colonises one or more woody species, whatever the context. Each of them usually has specific requirements, as insects do not attack any wood under any conditions: softwood or hardwood, coniferous or deciduous wood or both, standing wood, dead wood, decaying wood or storage wood, in the presence of fungi or not, etc. Each one should therefore be studied on a case-by-case basis. To sum up, infestation can occur at the following times, ranked from the most ancient to the most recent one: when the wood is standing in full vigour, when it is diseased or freshly felled/dead, during a possible storage phase, when it is used as timber or as an object, when it is in advanced decomposition. In archaeology, infestation can be recorded after burying the remains and before archaeological investigations between two excavation campaigns or during excavation (if it rains a lot, for example, or if conditioning is bad), or even after the wooden remains have been conditioned. Then comes the heritage area with infestations during exhibitions or during the storage of works in collections.

This book is thus devoted to insects with a xylophagous diet, particularly in their larval state, more rarely in the adult state, as well as to some insects that perforate wood in order to nest (carpenter bees, ants, termites). Woodborers are insects whose larvae develop in woody plants and mainly belong to four main orders: beetles (capricorns, Ptinidae, bark beetles, weevils …), isopterans (termites), hymenopterans (*sirex*) and certain diptera such as the wood-cake beetle for example. In fact, all the insects that degrade wood and thus affect its mechanical properties are concerned. However, this excludes live-wood pests such as biting-sucking insects, phytophagous insects and most saproxylophagous insects such as chafers, which are more often found in litter or compost. A few other xylophagous insects belonging to beetles (such as certain buprestes, platy-pus) but also belonging to Lepidoptera (*Cossus cossus, Zeuzerina pyrina*, some Sessidae …) could not be studied. Consequently, they are not mentioned in this work and will be the subject of a later addition. Finally, marine woodborers, which are not insects but molluscs such as *Teredo navalis*, whose biology and ravages could be the subject of specialised and independent work, were also excluded.

The present work is primarily a practical tool to assist in the work of determining the indirect traces left by insect pests, i.e., galleries and frass. For this reason, it is largely based on the presentation of various anatomical sections of infested wood combined with the description of the indirect traces left by wood-boring insect larvae on their feeding material. Since its aim is to help preserve forests, buildings or objects, whether heritage ones or not, it is a means of gaining a better understanding and assessing the condition of a wood, structure, tree, forest, etc., and even of being able to reconstruct the environment and/or the context that led to this condition. This way, it helps in writing the history of the construction and deconstruction of buildings.

Because they are perfectly integrated into the forest ecosystem or the timber, it is essential to know the main biological characteristics of these insects (their diet, reproduction cycle, mode of action) in order to better prevent their multiplication and control them.

In addition, it is necessary to be aware of the anatomical diversity of wood species and insects and, above all, of the adapt-ability of living organisms to different habitats and environmental or climatic conditions, which offers thousands of possible combinations that cannot all be described here.

The identification key is accompanied by an atlas illustrating some of the various families, genera and species of wood-boring insects. This book does not claim to be exhaustive, but is intended to be a workbook for the reader, who should take the potential of such a subject, its variability and therefore its complexity into consideration.

References

Berger P., 2012. *Coléoptères Cerambycidae de la faune de France continentale et de Corse*. Actualisation de l'ouvrage d'André Villiers, 1978. Ed. Association Roussillonaise d'entomologie (A.R.E.), Perpignan, 664 p.

Blanchette R. A., 1991. *"Deterioration in historic and archaeological woods from terrestrial sites"*. *In*: Koestler R.J., Koestler V.H., Charola A.E., Nieto-Fernandez F.E. (eds) *Art, biology, and conservation: biodeterioration of works of art*. The Metropolitan Museum of Art, New York, NY, pp. 328-347.

Bobadilla L., Arriaga F., Luengo E., Martinez R., 2015. "Dimensional and morphological analysis of the detritus from six European wood boring insects". *Maderas, Ciencia y tecnologia*, 17 (4), pp. 893-904.

Brazier J., Franklin G. L., 1961. "Identification of hardwoods: a microscope key". *Forest products research*, No. 46. Department of Scientific and Industrial Research, London, 96 p.

Bremond J., Lessertisseur J., 1973. « Lamarck et l'entomologie ». *Revue d'histoire des sciences*, 26 (3), pp. 231-250.

Centre Technique du Bois et de l'Ameublement, 1996. *Le traitement curatif des bois dans la construction*. Département BIOTEC (Biologie, Environnement, Technologies). Eyrolles, Paris, 140 p.

Couvert M., 1977. *Atlas d'anatomie des charbons de foyers préhistoriques. Afrique du Nord tempérée, Mémoires du centre de recherches anthro-pologiques préhistorique et ethnographiques*. Alger, SNED, Mémoires du CRAPE, vol. 26, 28 p.

Español F., 1992. *Coleoptera, Anobiidae. Fauna Iberica*. Ed. Madrid, coll. Departamento de Biologia Animal, Universitad de Barcelona, Museo Nacional de Ciencas Naturales. Consejo Superior de investigaciones cientificas, 196 p.

Gambetta A., 2010. *Funghi e insetti nel legno*. Diagnosi, prevenzione, controllo (Italian) Paperback, Libro universitario. Nardini Editore, Florence, 155 p.

Greguss P., 1955. *Identification of living gymnosperms on the basis of xylotomy.* Akademiai Kiado, Budapest, 263 p.

Hoffmann A., 1986. *Faune de France, 59, Coléoptères, Curculionides (Deuxième partie).* Ed. de 1954 réimprimée, éd. Fédération française des sociétés de sciences naturelles, Paris, 1211 p.

Huber B., Rouschal C., 1954. *Mikrophotographischer Atlas mediterraner Hölzer.* Fritz Haller, Berlin, 105 p.

Jacquiot C., Trenard Y., Dirol D., Boureau M. E., 1973. *Atlas d'anatomie des bois, les Angiospermes.* Centre technique du bois, Paris, 75 p.

Lepesme P., 1944. *Les coléoptères, des denrées alimentaires et des produits industriels entreposés.* Paul Lechevalier, Paris, 124 p., 233 fig., 12 pl.

Pitman A. J., Jones A. M., Gareth Jones E. B., 1993. "The wharf-borer *Nacerdes Melanura* L., a threat to stored archaeological timbers". *Stud. Conserv.*, 38, pp. 274-284.

Ruffinatto F., Crivellaro A., 2019. *Atlas of macroscopic wood identification with a special focus on timbers used in Europe and CITES-listed species.* Springer Nature Switzerland AG, Cham., 439 p. https://doi.org/10.1007/978-3-030-23566-6

Schweingruber F. H., 1990. *Anatomie europäischer Hölzer; anatomy of European woods.* WSL-FNP. Haupt, Birmensdorf, 800 p.

Vernet J.-L., coll. Ogereau P., Figueral I., Machado Zanes C., Uzquiano P., 2001. *Guide d'identification des charbons de bois préhistoriques et récents, Sud-Ouest de l'Europe: France, péninsule ibérique, et îles Canaries.* CNRS Editions, Paris, 395 p.

Chapter 1
Presentation and Work Guide

1.1 Starting Point

The starting point of this work is a study of archaeological carbonised wood. The entomological remains had not been preserved by the burial of the remains, whereas the wood had, thus limiting interpretations. Wood, although rare in archaeology, is only preserved in certain forms and in certain specific cases: charcoal (following fires), silicified wood (in the desert), waterlogged wood (in humid, sub-aquatic contexts …). But insect galleries and frass can also be preserved (Fig. 1.1). It was during these studies that observations of insect traces on existing woods began. This led to the finding that these traces were family- or even species-specific, and that the condition of the wood varied very little in terms of macroscopic and microscopic measurements and characteristics. This fact was confirmed by carbonisation experiments on infested wood under different hygrometric and thermal conditions (Fig. 1.2). The characterisation effort initially focused on the insects most frequently encountered in dry and worked wood such as *Anobium punctatum* and *Hylotrupes bajulus*. Then, in a second step, pests of standing and dead forest wood were observed. Finally, the bioecology was found species-specific.

The observations of insect traces recorded in the atlas come from infested wood collected during harvesting missions in the forest or resulting from laboratory rearing. Given the high number of European wood-boring insects, choices had to be made. Obviously enough, the species most frequently encountered during our investigations were selected. But this tool is flexible: the possibility of adding new species remains open as this is a long-term task that has only just begun.

The construction of this atlas lies at the interface of animal and plant biology. The traces of woodworm larvae on wood, and more broadly the pathologies of wooden structures, are an increasingly important concern for the building sector, but also for heritage conservation and archaeology. It is therefore essential to explain the codes used in the present work.

Thus, the description of xylophagous insects is based on the current international entomological nomenclature, i.e., the nomenclature used for the construction of taxonomic keys. Updates in the names of certain families, genera or species, such as Anobidae now called Ptinidae, have been taken into account. Species are designated using Latin terms in order to avoid confusion. A same common name is indeed sometimes given to several species, even though they are different, e.g., for Ptinidae, whose subfamily Anobiinae includes 108 species. Or again, it is a question of avoiding confusions in the taxa of the former identification key that has been revised since: for example, the "Bostryche typographer", from the Latin *Ips typographus*, actually belongs to the family Scolytinae and not to the family Bostrichidae. Likewise, references to woody species and the anatomical elements of plants are based on the vocabulary generally used in biology, xylology and bioarchaeology.

1.2 A Few Biological Elements

1.2.1 Various Types of Pests

Four main types of wood pests are considered in this book:

- Corticophagous pests live on or under the more or less dehiscent bark of healthy, diseased or dead (standing or felled) trees and shrubs. The rather cambiophagous larvae (see below) colonise stumps and recently cut roots. Only the adult is considered to be corticophagous, notably because it feeds on the bark of young stems and branches, most often from outside the tree, causing small wounds to the subject (e.g., *Hylobius* spp., some tenebrionids, psocids …). Entire planta-

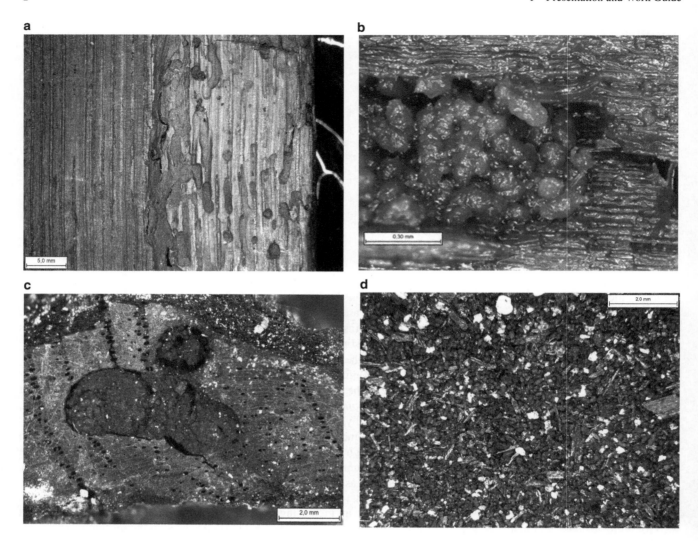

Fig. 1.1 **a** Archaeological waterlogged wood with *Anobium* infestation, galleries. **b** Archaeological waterlogged wood with *Anobium* infestation, faecal pellets. **c** Galleries on medieval wood charcoal. **d** Frass composed by faecal pellet carbonised, preserved on ancient wood charcoal

tions can be decimated in case of an outbreak. This type of pest is to be differentiated from biting-sucking insects (aphids, scale insects: Homoptera) which are not dealt with in this book (Table 1.1).

- Cambiophagous pests are specifically sub-cortical insects whose larvae feed exclusively on the cambium of the tree, destroying the wood-generating base of the trunk and branches. These insects are among the most formidable parasites of forest trees. They usually attack weakened trees, but their development can sometimes become epidemic (e.g., *Ips* spp., *Scolytus* spp., *Tomicus* spp., (Scolytinae); *Pissodes* (Curculioninae); *Agrilus* spp., *Melanophila cyanea* (Buprestidae); *Dioryctria sylvestrella* (Lepidoptera, Pyralidae)).

- Xylemophagous pests represent the largest group of woodborers; these insects feed on the whole wood, from the bark to the sapwood, passing through the cambium, and sometimes penetrating the heartwood. They infest dry wood and fresh or dead wood (e.g., *Saperda* spp. (Cerambycidae); *Cryptorhynchus* (Curculioninae); some Bostrichidae, Buprestidae, Lyctidae, Ptinidae, some Scolytinae; *Xyleborus*, *Platypus*; *Sirex* spp. (Hymenoptera, Siricidae); termites (Isoptera); *Zeuzera* spp., *Cossus cossus* (Lepidoptera, Cossidae)).

- Saproxylophagous pests are insects and other organisms that infest wood whose decomposition stage is advanced (worm-eaten, spongy, powdery wood) generally by the above-mentioned xylemophagous insects, mostly in association with lignivorous fungi. This is an important part of biodiversity (e.g., Cetoniidae, some Curculionidae, Melolonthinae …).

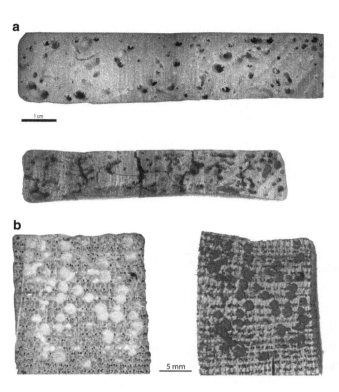

Fig. 1.2 a Experimentation before and after carbonisation (*Oligomerus* infestation). **b** Experimentation before and after carbonisation (*Lyctus* infestation)

Table 1.1 Frequency rating of the type of pest according to the condition of the wood. x: infrequent; xxx: very frequent

	Diseased tree	Felled/logged wood	Dry wood/timber	Damp wood/dead wood	Decayed wood
Corticophagous	xx				
Cambiophagous	xx	xx	x		
Xylèmophagous	x	xx	xxx	xx	x
Saproxylophagous				xx	xxx

The biology of some families, genera or even species is still poorly understood. On the one hand, a large part of the existing bibliography mainly corresponds to systematics, i.e., the description of the different anatomical parts of adult insects and larvae: elytra, genitalia, mandibles, antennae, bristles … On the other hand, some species, although found in large numbers in a given place, region or forest, remain rare for a given territory as a whole, and their habits are difficult to observe. In certain rare but possible cases, problems arising from the systematic approach, sometimes since the beginning of the history of the disciplinary field, have led to confusion between several species and their biology. Methods for observing and identifying insects have made great progress over the last 40 years, particularly through the use of new technologies such as molecular biology and insect DNA sequencing. However, there is still a tangible knowledge gap about the habits and bio-ecology of woodborers. Certain re-observations made in the 1940s–1950s, themselves based on old observations sometimes dating back to the nineteenth century (1870s) have to be taken into consideration, as well as certain confusions or "inaccuracies" in the anatomical descriptions of the individuals or their bio-ecology. Little by little, current tools and research are tending to update these states of knowledge. Finally, the sometimes very important adaptability of a species to its environment must be taken into account. This makes it difficult to give an exhaustive description of all its habits and preferences.

1.2.2 Wood Composition: What a Mixture!

Xylophagous insects do not attack any wood under any conditions. In general, three conditions favour colonisation: temperature, wood humidity and wood nutritional value. Thus, depending on their nutritional needs, wood-boring insects and more particularly beetles infest living wood, or dead and more or less dry wood, or wood colonised by fungi … Just like insects, each wood species has its own chemical composition (mineral elements and organic substances) with concentrations sometimes varying among individuals depending on the environment (shaded slopes, thermophilic species, calcareous soils, tropical woods …). Each insect adapts, chooses and digests some of these elements according to its digestive juices (diastases, enzymes) and its nutritional needs (cellulose, starch …). The digestive system of insects feeding on fresh wood is more complex than that of insects feeding on dry wood.

Below are the main feeding characteristics selected by insects, common to all woods but in different concentrations depending on the wood:

- 50% cellulose, a major, hydrophilic, polysaccharide-type (sugar) constituent of cell walls
- 25% hemicellulose, more complex polysaccharides
- 25% lignin, a hydrophobic constituent of the secondary wall of wood cells (phenylpropane, alcohols).

These three macromolecular substances include attractive or repellent extractables depending on the wood-eating insects:

- mineral salts
- starch
- reducing/soluble sugars (e.g., glucose, fructose, sucrose)
- proteins (amino acids, peptides) (1–2% in dead wood)
- tannins
- oleoresins
- oils and fats
- organic acids
- organic nitrogen compounds.

For example, Lyctidae and Bostrichidae feed on intermediate polysaccharides such as starch and hemicelluloses. Digestion by Scolytidae is more complex, as they assimilate polysaccharides from the cell content, but also from the cell walls, including hemicelluloses and wall carbohydrates (other than cellulose). Ptinidae and Cerambycidae use all the polysaccharides of the cell walls and cell contents (proteins and lipids) as well as celluloses. Buprestidae seem to favour hemicelluloses and xyloses. Finally, saprophagous insects such as chafers (genus *Cetonia*) or rhinoceros beetles (genus *Oryctes*) favour cellulose and pentosans.

As each species has a specific biology, the consequences of infestation on wood health are variable, all the more so as these biological alterations are not the only ones to interplay (fungi, pesticide treatments, the environment of the structure, its age …, also play a role). Generally speaking, biological alteration alters the mechanical and structural qualities of the wood, making the whole structure much more fragile under the load it is supposed to carry, until the most important parts break. Two different species of insects will take different lengths of time for this stage to be reached. For example, infestation by capricorn beetles will be much more destructive in the short term than infestation by Ptinidae.

1.2.3 The Life Cycle of a Wood-Boring Beetle

The life cycle of wood-boring beetles is simple. It includes the time following egg laying, the larval stages, pupation, and transformation into an adult. This time varies across species, ranging from a few months to several years. In fact, the egg, deposited in a gallery or in a crack in the wood, has a lifespan of a few days. After hatching, the larvae move into the wood and feed on it, causing damage. The larvae evolve through successive moults called "larval stages". The duration of each larval stage depends on the environmental conditions and the nutritional quality of the wood: the more favourable the conditions, the faster the cycle will be; otherwise, the larva may remain dormant for several months or even years. The pupal stage lasts a few weeks and corresponds to a phase of transformation into an adult during which the larva no longer feeds. Finally, the imago comes out. It does not feed, only lives a few weeks, and its function is essentially related to reproduction. However, in certain families such as bark beetles, the female bores a gallery in the wood to lay her eggs. Similarly, most weevils, once adult, continue to live and feed on wood before laying their eggs (Fig. 1.3).

The issue of the duration of an infestation is a regular concern for heritage officers in museums and the judiciary. For the latter, these questions arise during a dispute after the purchase of an infested property, most often caused by the merula—a lignivorous fungus—or a dry-wood-boring insect. An expert then tries to deduce "a first date" from the infestation state in order to tell who should pay for the repair work—the former owner or the new one. The age of the frass on old or recent wood, church or house frames, art objects, furniture, etc., is a complex point to apprehend. At present, it is impossible to determine with certainty whether the frass visible on a structure is recent or whether it has been present for several years. The hypothesis that the frass is compressed and detaches itself "en bloc" from the gallery would be an indication of age, as

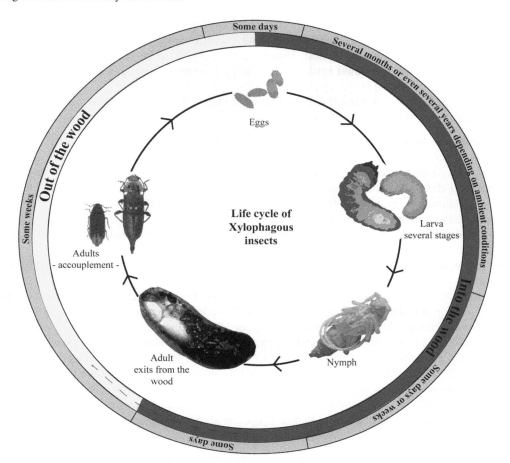

Fig. 1.3 Development cycle of wood-boring insect, coleoptera

opposed to "light and volatile" frass. However, no precise study has yet been carried out on the subject. DNA analysis and analysis of amino acids and other enzymes used in insect digestion are ways of answering this question. But everything still remains to be developed.

For the frass preserved in archaeological woods, the question remains the same when it comes to waterlogged wood taken from ancient excavations and dried naturally, increasing its attractiveness to today's insects. On the other hand, in the case of waterlogged wood from recent excavations, which has been removed and properly conditioned to keep it in a good state, insect traces can be considered to be original, i.e., dating back to before the remains were buried. The case is simpler for carbonised wood because no insect can feed on carbonised wood. Thus, when a charred structure is discovered, the preserved galleries and frass attest to an infestation before the charring event and probably at the time the structure was used.

The duration of colonisation by xylophagous insects can be more easily apprehended based on knowledge of the insect's biology, i.e., the length of the development cycle and the number of generations observable in the woods. This known period of time would open up new prospects for interpretation. For the moment, however, it is impossible to define an exact number of generations, as this number depends on the species in question and on the ambient environment (temperature, humidity, etc.). If conditions are optimal, generations follow one another more quickly than in difficult conditions (harsh winter, too dry weather, etc.). Data about the development cycle of each species are therefore only indicative and do not constitute absolute truth.

1.3   Guide to Using the Identification Key and the Atlas

1.3.1   Useful Points

The traces of xylophagous insect activity were identified, in parallel with the identification of the woody species, on the basis of three main criteria: the host, the gallery, and the frass.

The predominance of one or the other character in the samples made it necessary to develop two distinct identification keys: one for the galleries and frass, and one for the frass alone. Galleries are the most easily observable and common characteristic. They can be filled with frass or sawdust. However, it is possible to harvest the frass outside the context of the source gallery, especially when it is rejected outside the wood by the larvae. Finally, knowledge of the feeding host may exclude certain species.

Depending on the work context and disciplinary field, as much data as possible should be collected on the sample before starting to read the key.

- **The first criterion to be taken into account concerns the host, i.e., the feeding medium.**
  Angiosperm, gymnosperm or other
  Some insects are polyphagous, but others are inferred from a particular woody species or group that depends on an anatomical or nutritional characteristic of the tree.
  Is the support a deciduous or a coniferous tree? Or a monocotyledonous plant?
  Is it a soft wood such as poplar (*Populus*) or willow (*Salix*) for example?
  Is it a wood with large vessels (vessels of the earlywood) such as oak (*Quercus*), chestnut (*Castanea*), fig (*Ficus*)?
  Host part
  As mentioned above, some insects will prefer one or more parts of the host: branches, crown, trunk, roots ….
  State of the wood
  Depending on the case, it is possible to know the state of the wood before the identification of insect traces, while in other cases, as in archaeology, it is the identification of insect traces that helps to grasp this aspect.
  Is it a living tree in full vigor? A dead tree, still standing, or fallen to the ground? A freshly felled tree that has not been debarked? Is it wet/dry? Is it worked wood? To what extent are the mechanical properties of the wood impaired?
  Did the wood react to the infestation?
  All the wood cells affected by a wood borer's attack should be observed. When dead, stored or worked wood is infested, not all the cells are deformed before and after the infestation. In contrast, during infestation of standing wood, the wood cells continue to grow and try to repair the damage caused by the insect or parasite. In this case, deformations of the wood anatomy are observed, with scarring cells and rings that shift the arrangement of the wood cells in the infested area.

- **The true first characteristic marker of a family of xylophagous insects is the gallery.** Observed alone, it does not provide enough information to determine the species. Priority is given to the analysis of exit galleries or flight holes, i.e., the galleries bored and used by adult individuals just after nymphosis to get out of the wood and reproduce.
  Shape and diameter
  The general shape of the gallery is an important determination criterion, together with the size of the flight hole. For diameter measurements, insect flight holes are preferred because they are closer to the maximum size of the species.
  The position of the gallery in the wood
  The gallery should be localised within the anatomy of the wood: under the bark only, under the bark and plunging into the sapwood, specifically in the sapwood, the heartwood, or in the entire wood.
  The targeting of specific wood cells
  Depending on the species, the larvae may prefer certain wood cells rich in nutrients such as starch, that are necessary for their development. Pests then focus on only those cells with a high parenchyma content, such as the cells of the earlywood. Others do not seem to distinguish between the different parts of the wood (rays, initial/final wood, etc.).
  Gallery arrangement
  Depending on the species or more broadly on the families, it is possible to observe a certain organisation in the layout of galleries. For example, they may be parallel to each other, starting all from the same point of origin (a mother's gallery, for example).
  Sometimes the galleries are arranged in different ways, which can be seen in relation to the grain of the wood (parallel, oblique, perpendicular). In general, wood-boring larvae generally avoid competition with their congeners. Crossing galleries can be observed in case of overpopulation or several successive generations. In this case, the galleries intersect in an anarchic manner ("networks").

- **The second milestone in the identification of woodworm traces is the frass.** Shapes vary from one species to another, but the same type of shape is generally found *per* family. The frass (or excrement) of wood-boring insect larvae is observed from the largest to the smallest scale. Apart from its presence or absence, not to be confused with the wood chips produced by some wood-boring insects, we must first try to determine whether it is homogeneous or not; then characterise the shape of the faecal pellet that composes this frass, which is often different for each family, or even each species on a microscopic scale.

Presence or absence of frass

The observation of a gallery does not necessarily induce the presence of frass. In fact, the presence or absence of frass may indicate the presence of a particular group or family of xylophagous insects.

However, we should be careful with archaeological wood:

- Charcoal may vitrify to a higher or lesser degree during carbonisation. Frass can also be deformed by this phenomenon.
- It is important to know that frass is not easily preserved in waterlogged wood, where it is also waterlogged and may therefore swell and deform more easily. It is necessary to dry the sample to the critical point (dehydration by progressive dilutions with ethanol) to restore the original aspect of the frass, which becomes identifiable in this way. However, this process does not always produce reliable results, and measurements made in this way must be treated with caution because of the margin of error.

General aspect of the frass

The homogeneity or heterogeneity of the frass should be observed, i.e., whether the frass as a whole is only made up of individualisable faecal pellets (larval excreta with a particular shape depending on the species), or whether it is made up of various debris, excreta, pellets and more or less digested debris.

Homogeneous frass is composed of small elements of the same type and shape, called pellets. These pellets can also correspond to a set of digested wood debris without any particular agglomerate.

Heterogeneous frass is made up of elements that vary according to the frass type: wood fibres, digested wood agglomerates, powders of different types within the same gallery ….

Shape and size of a faecal pellet

Measurements of length and width or diameter can discriminate between a group of species or even a particular species. It is also essential to examine the general shape of a unit (spherical, cylindrical, fusiform, variable …) and to observe its occurrence within the whole frass.

Microscopic characteristics

A set of photographs of the faecal pellets, taken with a scanning electron microscope, brings to light details that may link them to a given species. The main ones are listed below:

- pellets made up of various digested wood debris arranged differently according to the species: parallel, anarchic, layered, smooth ….
- pellets displaying distinct features: a hollow or streaked line, a sharp apex, a rounded or curved base, a spiral apex ….
- the presence within the pellet of undigested woody cells (perforations, areolated punctuations, vessels, etc.).

- **Other clues are interesting such as the presence of fungi** (hyphae, cubic type rot …), pupal shells, seeds, etc.; they can be spotted during frass observation and provide valuable identification clues.

1.3.2 Decoding Atlas Data

The key described in Chap. 2 is more particularly composed of two distinct keys: the first one is useful if the sample to be analysed is complete (wood, gallery and frass), the second one is essential when the sample contains only frass with very little or no gallery, e.g., in certain heritage conservation diagnoses or in archaeology with often degraded galleries on the surface of burnt wood.

The key is also presented in the form of a diagram (Sect. 2.3).

The resulting atlas contains a variety of information that can be used by botanists, entomologists, archaeologists-botanists, etc. It is currently composed of several large groups of xylophagous organisms: Coleoptera, Hymenoptera and Isoptera. These large groups are divided according to large families in alphabetical order (Bostrichidae, Cerambycidae …). This is followed by a general description of the common biological characteristics of each family. Then a table of the most common species of this family in Western Europe is generally proposed. It brings together various data on the insect's food preferences and biology, when known. This data will certainly need to be updated according to future discoveries. It is followed by an individual presentation sheet for each species listed in the key. The sheet provides a brief description of the adult insect, its colour and size in particular. The current geographical distribution and possible origin of the species are given as an indication if the species is exogenous to Europe. Then, the bio-ecology of the insect is presented as well as its preferences, the infested wood, the minimum/maximum temperature and humidity when these indications are known. Finally, some indications on its development cycle can be provided.

Insects are included in each sheet; they enable the reader to spot the preferred hosts and the infestation type at a glance (Figs. 1.4 and 1.5).

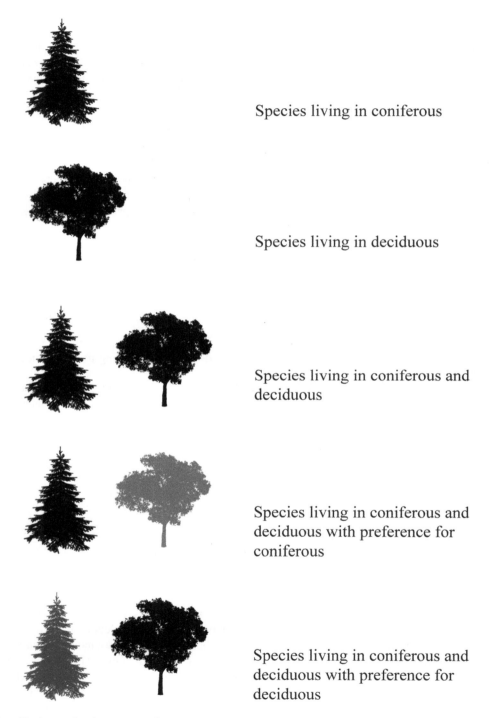

Species living in coniferous

Species living in deciduous

Species living in coniferous and deciduous

Species living in coniferous and deciduous with preference for coniferous

Species living in coniferous and deciduous with preference for deciduous

Fig. 1.4 Legend found in the species sheets concerning the type of infested tree

Fig. 1.5 Legend to be found
in the species sheets
concerning the type of
infestation

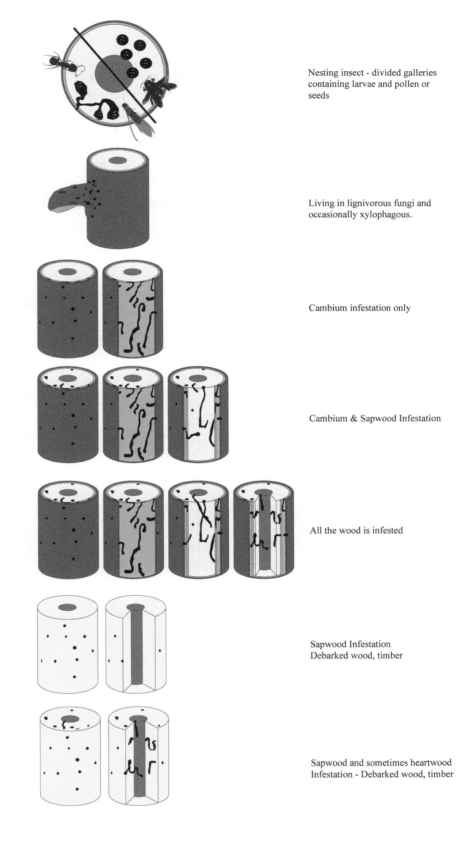

Nesting insect - divided galleries
containing larvae and pollen or
seeds

Living in lignivorous fungi and
occasionally xylophagous.

Cambium infestation only

Cambium & Sapwood Infestation

All the wood is infested

Sapwood Infestation
Debarked wood, timber

Sapwood and sometimes heartwood
Infestation - Debarked wood, timber

Each species sheet is accompanied by a series of illustrations:

Macroscopic characteristics

Photographs taken with a binocular microscope show galleries, which are the first visible signs of infestation. General views are followed by increasingly detailed views. The same is true for frass when it is present for the given species.

A description of the specific characteristics of each element (gallery and frass) can be found under each series of photographs.

Microscopic characteristics

These are SEM images enriching the previous illustrations with details invisible or difficult to see with a binocular microscope, especially as regards the faecal pellets composing the frass.

Drawings

A drawing board of the faecal pellets can also be added. These are schematic representations of the pellets aimed at showing the range of their variability.

Chapter 2
General Identification Keys

M. Toriti et al., *Traces of Common Xylophagous Insects in Wood*, https://doi.org/10.1007/978-3-030-66391-9_2

2.1   Key 1: Identification Key of the Galleries and Frass of a Few Xylophagous Insects

1
- Galleries or flight holes oval to elliptical, ≤3 cm in diameter, often 1–2 cm (Fig. 1)..**2**
- Galleries or flight holes oval to elliptical, large (>3 cm in diameter), on deciduous species only, in the cambium and the sapwood (Fig. 1)..***Cerambyx cerdo, C. welensii***
- Exit holes of different shapes and sizes (Figs. 2 and 3) **3**

2(1)
- Galleries oval to elliptical, located below the bark, in the cambium only (Fig. 4)................................**4**
- Galleries oval to elliptical, located below the bark, in the cambium and the sapwood (Fig. 5)............**5**
- Galleries oval to elliptical, located below the bark, in the cambium, the sapwood and the heartwood; in stumps and dead wood on the ground, in forests (Fig. 6) **6**
- Galleries oval to elliptical, generally located in the sapwood, sometimes also in the heartwood, on barked and worked woods..**7**

3(1)
- Galleries round, perfectly circular (Fig. 2)...**8**
- Galleries of an irregular shape, ovoid to round (Fig. 3) **9**

4(2)
- Type-1 Cerambycidae, primary pest, cambiophagous, ~20 species; on deciduous species only .... **Genera *Ropalopus, Saperda, Trichoferus***
- Type-1 Cerambycidae, primary pest, cambiophagous, on coniferous species only .... **Genus *Nothorhina***

5(2)
- Type-2 Cerambycidae, secondary pest, xylemophagous, on deciduous species only, ~40 species, including ....***Xylotrechus stebbingi***
- Type-2 Cerambycidae, secondary pest, xylemophagous, on coniferous species only, ~40 species....................**10**
- Type-2 Cerambycidae, secondary pest, xylemophagous, on deciduous and coniferous species, ~12 species, including ....***Penichroa fasciata***

6(2)
- Type-3 Cerambycidae, in the cambium through to the heartwood, sometimes on decomposing wood, deciduous species only, 18 species .... **Genera *Anoplodera, Clytus, Stenurella***
- Type-3 Cerambycidae, in the cambium through to the heartwood, sometimes on decomposing wood, coniferous species only, 7 species .... **Genera *Acmaeops, Asenum, Molorchus***
- Type-3 Cerambycidae, in the cambium through to the heartwood, sometimes on decomposing wood, deciduous and coniferous species, 16 species .... **Genus *Rhagium***

7(2)
- Type-4 Cerambycidae, in dry and/or worked woods, coniferous species only, 5 species, including .... **Genus *Arhopalus***
- Type-4 Cerambycidae, in dry and/or worked woods, deciduous and coniferous species, 4 species, including ....***Hylotrupes bajulus***

8(3)
- Exit gallery >3 mm in diameter..**11**
- Exit gallery <3 mm in diameter..**12**

9(3)
- Absence of frass (Fig. 7)..**13**
- Presence of frass (Fig. 8)...**14**

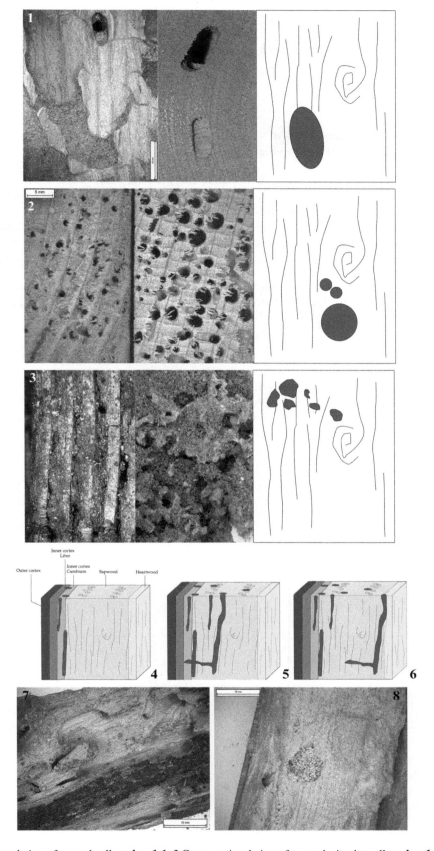

Figs. 1–8   1 Cross-sectional view of an oval gallery: **key 1:1**; 2 Cross-sectional view of a round, circular gallery: **key 1:3**; 3 Cross-sectional view of an irregularly shaped gallery: **key 1:1–3**; 4 Galleries in the cambium only: **key 1:2**; 5 Galleries in the cambium and the sapwood: **key 1:2**; 6 Galleries in all parts of the wood: **key 1:2**; 7 Frass-less gallery: **key 1:9–11**; 8 Gallery filled with frass: **key 1:9–11**

10(5) • Heterogeneous frass composed of half of wood chips and half of faecal pellets that is fine cylinders or "barrels" (Fig. 9) ....*Icosium tomentosum*

• Heterogeneous frass composed of >60% of wood chips and <40% of faecal pellets that is fine cylinders or "barrels" (Fig. 10)...*Callidium coriaceum*

11(8) • Absence of frass (Fig. 7)...**15**

• Presence of frass (Fig. 8) **16**

12(8) • Preferential infestation of the vessels of the initial wood and parenchyma zones (Fig. 11) ....**Lyctidae**

• No preferential infestation of the wood cells (Fig. 12)...**17**

13(9) • Very small galleries, <1 mm in diameter, presence of cubic rot or degraded wood (Fig. 13) ....**Acarids, saproxylophagous insects, wood decomposers**

• Galleries of variable sizes, whose presence has had an impact on tree growth, visible in growth rings (Fig. 14) ....**Biting/Sucking insects**

• Galleries of variable sizes, winding and forming networks, sometimes widening into "chambers", seeds sometimes stored inside (Fig. 15).......................................**Formicidae:** *Camponotus vagus*; **Rhinotermitidae**

14(9) • Galleries especially located below the bark, cambiophagous insect (Fig. 4).......................................**18**

• Galleries without any particular localisation relatively to the wood anatomy **19**

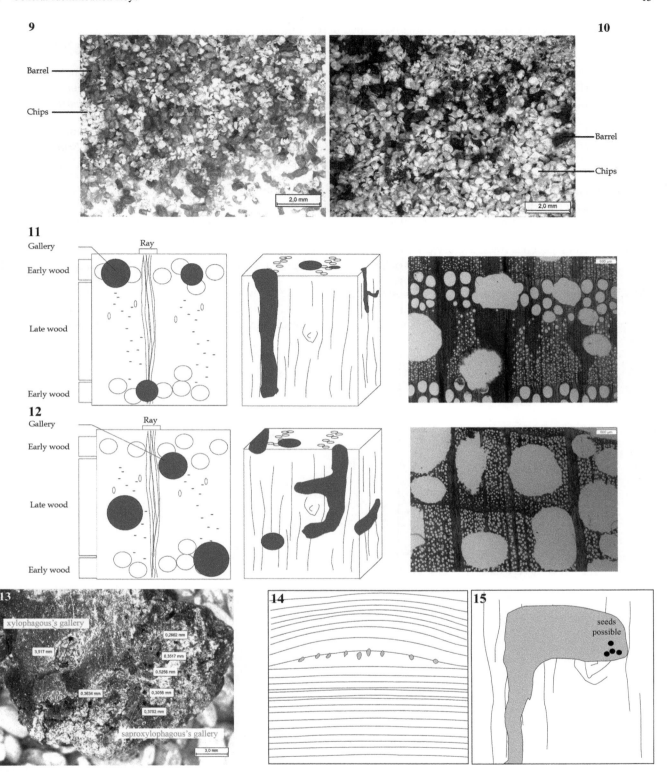

Figs. 9–15   9 Heterogeneous frass: wood chips (50%) and short cylinders (50%): **key 1:10; key 2:6**; 10 Heterogeneous frass: mostly wood chips (>60%) and short cylinders in a lesser quantity (~40%): **key 1:10; key 2:6**; 11 Preferentially infested cells: initial wood, rays … (cross-sectional view, 3d, and example under a thin blade): **key 1:12**; 12 Direction of the galleries independent of ligneous cells, variable relatively to the grain of the wood: parallel, perpendicular, oblique … (cross-sectional view, 3d, and example under a thin blade): **key 1:12**; 13 Galleries and degradation by saproxylophagous organisms, well decomposed wood: **key 1:13**; 14 Galleries inducing a reaction from the living tree: healing attempt: **key 1:13**; 15 Gallery growing wider into a cavity likely to contain seeds: **key 1:13**

15(11) • Presence of wood chips and straw-like, coiled fibres (Fig. 16) **20**

               • Calcareous envelope lining the inner side of the gallery (Fig. 17) .... **Mollusc:** *Teredo navalis*

16(11) • Frass granular, spherical, sometimes flattened or lentil-shaped or round, ~0.5–1 mm in diameter (Fig. 18) .... *Xestobium rufovillosum*

               • Fine frass, fine-flour type (Fig. 19) .... **Bostrichidae;** *Bostrichus capucinus*

17(12) • Faecal pellet granular, spherical, sometimes flattened or lentil-shaped to round, <0.5 mm in diameter (Fig. 20) .... *Ernobius mollis*

               • Fusiform faecal pellet (Fig. 21) **21**

18(14) • Galleries arranged in a geometrical pattern (Fig. 22), composed of a central primary gallery + several larval galleries on either side of the primary gallery, heterogeneous frass **22**

               • Galleries forming no particular pattern .... **Platypodinae; Lymexylinae; Buprestidae**

19(14) • Homogeneous frass, exclusively composed of pellets **23**

               • Heterogeneous frass, composed of pellets + various digested wood fragments **24**

20(15) • Galleries compartmentalised into several cavities made of wood chips; presence of pollen in each cavity (Fig. 23) .... *Xylocopa violacea, X. valga*

               • Galleries containing wood chips only or wood fibres, black bite .... **Lepidoptera,** *Sirex*

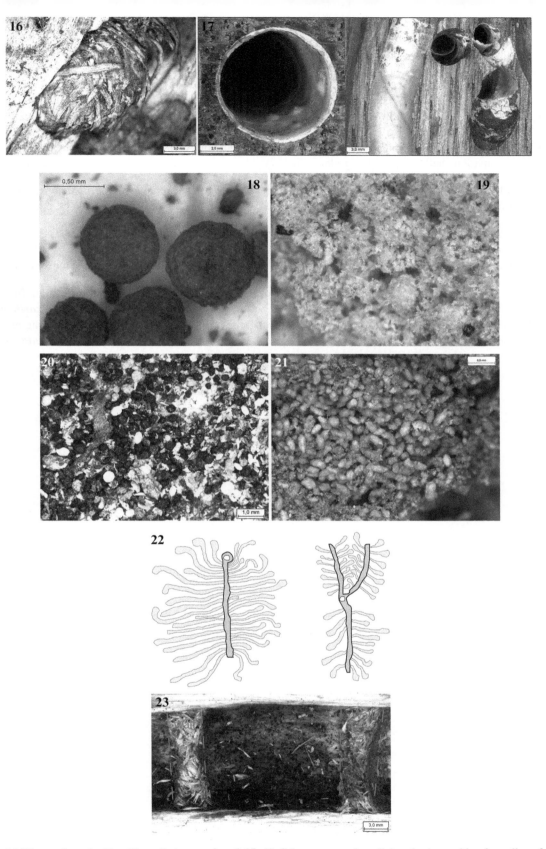

Figs. 16–23 16 Fibrous frass, looking like coiled straw: **key 1:15**; 17 Calcareous envelope lining the inner side of a gallery: **key 1:15**; 18 *Xestobium*-type frass, granular, lenticular: **key 1:16; key 2:4**; 19 Bostrichidae/Lyctidae-type frass, fine, powdery: **key 1:16; key 2:2**; 20 *Ernobius*-type frass, granular, lenticular: **key 1:17; key 2:4**; 21 Fusiform frass: **key 1:17; key 2:4**; 22 Galleries forming a geometric network, below the bark, composed of one or more central maternal gallery.ies + larval galleries around the initial gallery: **key 1:18**; 23 Gallery compartmentalised into cavities separated by wood chip walls, *Xylocopa* spp.: **key 1:20**

21(17) • Faecal pellet fusiform, short, rough, of variable shapes and sizes, altogether rounded with one rare sharp end, easily friable (Fig. 24) ....**Genus** *Anobium*
  • Faecal pellet fusiform, short to elongated, rough, regularly presenting a domed apex with one or two sharp ends (Fig. 25) ....**Genus** *Oligomerus*
  • Faecal pellet fusiform, short to elongated, rough, presenting a slightly curved apex with one sharp end, more rarely two sharp ends (Fig. 26)..**Genus** *Calymmaderus*
  • Faecal pellet fusiform, elongated and smooth, presenting a hollow ridge on its whole length and often two sharp ends, pupation shell composed of pellets (Fig. 27) ....**Genus** *Nicobium*

22(18) • Deciduous trees ....**Bark beetles of deciduous trees, e.g.,** *Scolytus multistriatus* **and** *S. rugulosus*
  • Coniferous trees...**Bark beetles of coniferous trees, e.g.,** *Ips typographus*

23(19) • Frass composed of faecal pellet that is concave seeds looking like maize grains (Fig. 28).................*Kalotermes*
  • Frass composed of shiny fusiform pellets, generally associated with a polypore fungus (Fig. 29) ....**Ciidae e.g.,** *Xylographus bostrichoides*

24(19) • Frass composed of half of partially digested wood fragments (fibres) and half of small fusiform pellets (Fig. 30) ....*Hexarthrum exiguum*
  • Frass mostly (70–80%) composed of small fusiform pellets, and of partially digested wood fragments (fibres) in a lesser quantity (20–30%) (Fig. 30) ....*Pentarthrum huttoni*

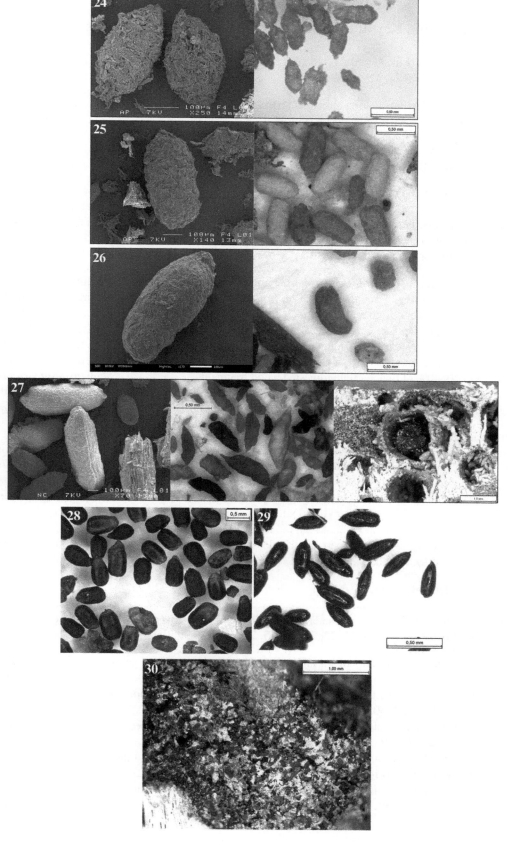

Figs. 24–30 24 Frass of *Anobium punctatum*: **key 1:20; key 2:8**; 25 Frass of *Oligomerus ptilinoides*: **key 1:21; key 2:8**; 26 Frass of *Calymnaderus solidus*: **key 1:21; key 2:8**; 27 Frass of *Nicobium castaneum*: **key 1:21; key 2:8**; 28 Frass of *Kalotermes flavicollis*, with pellets maize-grain-like: **key 1:23; key 2:4**; 29 Fusiform pellets of *Xylographus bostrichoides*: **key 1:23; key 2:8**; 30 Heterogeneous Cossoninae-type frass: **key 1:24; key 2:5**

2.2   Key 2: Identification Key of the Frass of a Few Xylophagous Insects

1   • Homogeneous frass (Fig. 31) **2**
           • Heterogeneous frass (Fig. 32)...**3**

2(1)   • Frass composed of pellets having a distinct shape **4**
           • Frass composed of powder only, absence of units (Fig. 19)..............................**Lyctidae/Bostrichidae**

3(1)   • Frass composed of pellets and fine, partially digested wood fragments **5**
           • Frass composed of short cylindrical pellets and wood chips..**6**

4(2)   • Pellets round-shaped, convex, or "maize-grain-like"
             (Figs. 28, 29, 30, 31, 32, and 33)..**Kalotermitidae**
           • Cylinder-shaped pellets, short cylinders or small "barrels" (Fig. 34) ....*Hylotrupes bajulus*
           • Pellets spherical, sometimes flattened, lenticular (Figs. 18, 19, and 20)..**7**
           • Pellets fusiform, or "peanut-like" (Fig. 21) **8**

5(3)   **Curculionidae family; subfamilies Cossosinae and Scolytinae (Figs. 30, 31, 32, and 33)**
           • Frass granular, powder-like, composed of half of very fine fusiform pellets (0.2 mm in length) and half of fine,
             partially digested wood fragments.. *Hexarthrum exiguum*
           • Frass granular, powder-like, composed of 70–80% of very fine fusiform pellets (0.2 mm in length) and 20–30%
             of fine, partially digested wood fragments .... *Pentarthrum huttoni*
           • Frass granular, often compacted, composed of 60–70% of very fine subspherical or even fusiform pellets (0.3 mm in
             length) and 30–40% of fine, partially digested wood fragments ....*Scolytus multistriatus, S. rugulosus*
           • Frass granular, often compacted, composed of half of very fine subspherical or even fusiform pellets (0.4 mm in
             length) and half of fine, partially digested wood fragments ....*Ips typographus*

6(3)   **Fresh-wood Cerambycidae family (Figs. 9, 10, and 33)**
           • Frass composed of ~20–30% of small, brittle "barrels" (<1 mm in length and 0.4 mm in width on average) and
             ~70–80% of small, convex chips (0.3–0.4 mm in width)...*Callidium coriaceum*
           • Frass composed of ~50–60% of small "barrels" (~1 mm in length and 0.6 mm in width on average) and 40–50%
             of small, convex, rather narrow chips (0.3 mm in width)..*Icosium tomentosum*
           • Frass composed of half of friable, cylindrical, easily compacted "barrels" and half of very fine chips (<0.3 mm in
             width) (40–50%)...*Penichroa fasciata*
           • Frass composed of ~40% of small, friable, cylindrical, "barrels" and 60% of numerous fine chips
             (<0.2 mm in width)..*Xylotrechus stebbingi*

7(4)   • Spherical pellets ≥0.5 mm in diameter (Figs. 18, 19, 20, 21, 22, 23, 24, 25, 26, 27, 28, 29, 30, 31, 32,
             and 33)..*Xestobium rufovillosum*
           • Spherical pellets <0.5 mm in diameter (usually 0.3 mm), with a groove visible on each unit; pupation shell
             composed of pellets (Figs. 20, 21, 22, 23, 24, 25, 26, 27, 28, 29, 30, 31, 32, and 33)...................*Ernobius mollis*

8(4)   • Pellets fusiform, short, rough, of variable shapes and sizes, rounded with one rare acute end, easily friable
             (Figs. 24, 25, 26, 27, 28, 29, 30, 31, 32, and 33)...*Anobium punctatum*
           • Pellets fusiform, short to elongated, rough, regularly presenting a domed apex whose end(s) can be sharp
             (Figs. 25, 26, 27, 28, 29, 30, 31, 32, and 33)..*Oligomerus ptilinoides*
           • Pellets fusiform, short to elongated, rough, regularly presenting a domed and curved apex whose end(s) can be
             sharp (Figs. 26, 27, 28, 29, 30, 31, 32, and 33) ....*Calymmaderus solidus*
           • Pellets fusiform, elongated and smooth, grooved along their whole length, often presenting two sharp ends, pupation
             shell composed of pellets (Figs. 27, 28, 29, 30, 31, 32, and 33).......................................*Nicobium castaneum*
           • Pellets fusiform, smooth and shiny, generally presenting one sharp, sometimes elongated apex, mycelium mixed
             with the frass (Figs. 29, 30, 31, 32, and 33) ....*Xylographus bostrichoides*

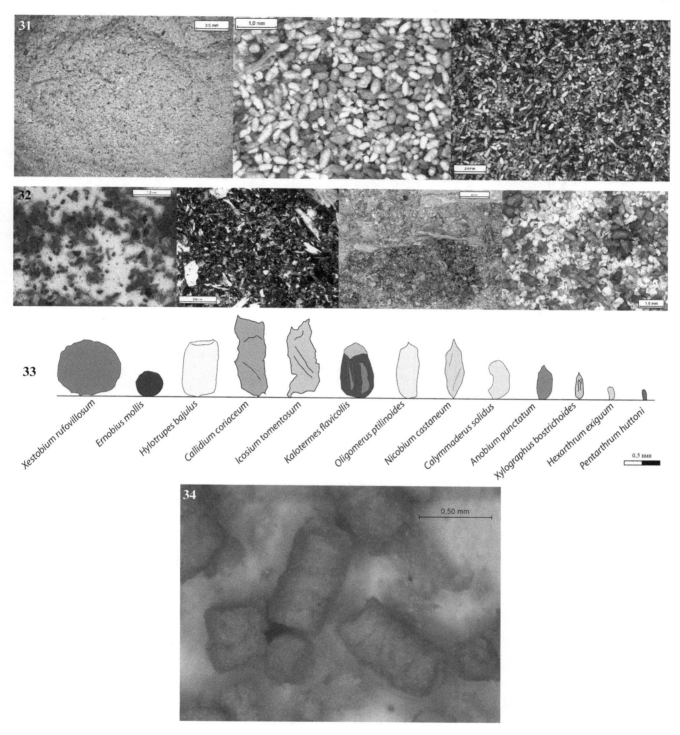

Figs. 31–34 31 Homogeneous frass: **key 2:1**; 32 Heterogeneous frass: **key 2:1**; 33 Schematic drawings of a few typical and specifically identifiable faecal pellets: **key 2:4–8**; 34 Small, short cylinders of *Hylotrupes bajulus* pellets: **key 2:4**

2.3   Key Diagrams

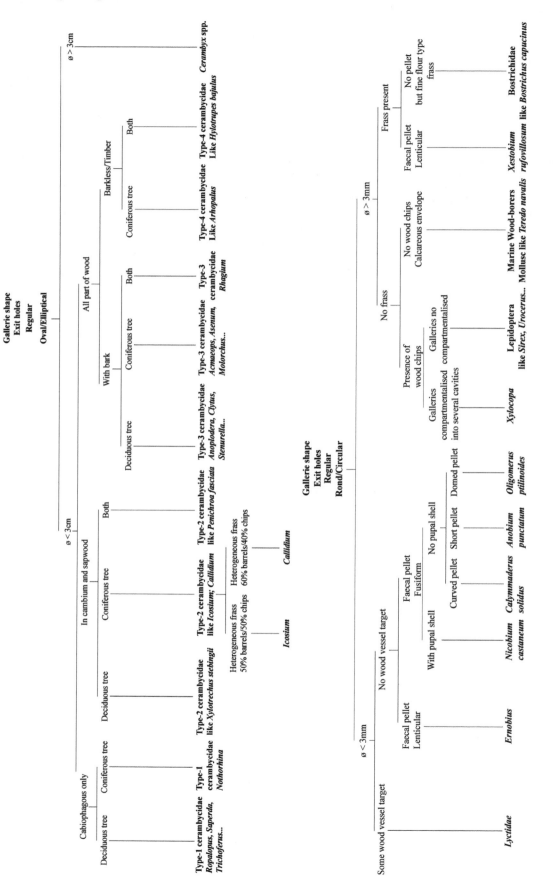

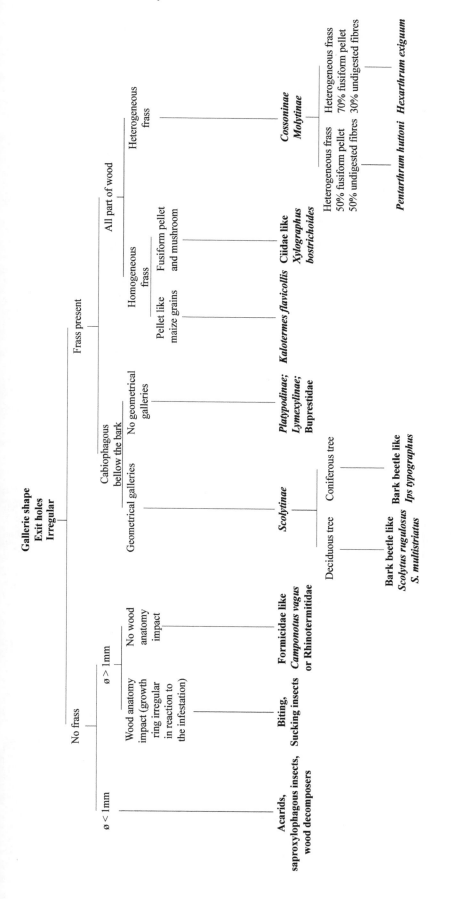

Gallerie shape
Exit holes
Irregular

No frass

ø < 1mm

Acarids,
saproxylophagous insects,
wood decomposers

ø > 1mm

Wood anatomy
impact (growth
ring irregular
in reaction to
the infestation)

Biting,
Sucking insects

No wood
anatomy
impact

Formicidae like
Camponotus vagus
or Rhinotermitidae

Frass present

Cabiophagous
bellow the bark

Geometrical galleries

Scolytinae

Deciduous tree

Bark beetle like
Scolytus rugulosus
S. multistriatus

Coniferous tree

Bark beetle like
Ips typographus

No geometrical
galleries

Platypodinae;
Lymexylinae;
Buprestidae

All part of wood

Homogeneous
frass

Pellet like
maize grains

Fusiform pellet
and mushroom

Kalotermes flavicollis   Ciidae like
Xylographus
bostrichoides

Heterogeneous
frass

Cossoninae
Molytinae

Heterogeneous frass
50% fusiform pellet
50% undigested fibres

Pentarthrum huttoni

Heterogeneous frass
70% fusiform pellet
30% undigested fibres

Hexarthrum exiguum

Chapter 3
Atlas of the Most Common Xylophagous Insects

3.1 Coleoptera

3.1.1 Bostrichidae

Presentation of the Family

Bostriches (or bostryches) form the family Bostrichidae that hosts more than 700 species divided into about 100 genera, most of which are xylophagous or even saproxylophagous. The greatest species diversity is found in the tropical and subtropical regions. These insects can cause significant damage to wood, especially timber. In Western Europe, around 30 species have been inventoried on wood or monocots.

This family has long been mistaken for and classified as a subfamily of Scolytidae (bark beetles), which are more commonly found on dying standing trees or freshly felled logs. Similarly, their biology is regularly compared with that of Lyctinae: this is why they are called false powderpost beetles. Moreover, the name "*Bostrychus typographus*", which is commonly used in the literature, is distinct from a member of the Lyctinae family known as the Engraver beetle from its true Latin name *Ips typographus*. *Ips/Bostrychus typographus* and bostriches belong to different families according to the current nomenclature. As for the typographer scolyte, it is inventoried within the family Scolytidae and subfamily Scolytinae.

Some species are exogenous (e.g., the genus *Dinoderus* or the species *Sinoxylon conigerum*): they originally came from Asia or Africa but are now acclimatised to our regions. Species growing in stored grains (the genus *Rhyzopertha*) such as maize and other cereals are not taken into account in this atlas. The majority of bostriches feed on deciduous trees, except for the genus *Stephanopachys*, which is strictly coniferous. Many species infest monocots like bamboo or reed (*Arundo* spp.) (*Bostrychoplites cornutus*; the genera *Dinoderus, Heterobostrychus, Scobicia*; *Micrapate xyloperthoides, Sinoxylon ruficorne, Xyloperthella picea*). Some species infest fruit trees (*Apate monachus, Scobicia chevrieri, S. pustulata*), or the Mediterranean flora (*Lichenophanes numida, Schistoceros bimaculatus*; the genus *Scobicia*; *Xylopertha praeusta*).

Bostriches usually spend their larval time in the cambium of bark and in the sapwood of branches and trunks of dying, injured or freshly cut and stored trees. Freshly cut and stored trees may have favoured the accidental transport of exogenous species from one country to another (e.g., *Sinoxylon* spp. on transport pallets and crates). *Bostrichus capucinus* and *Heterobostrychus aequalis* have also been reportedly found on tree stumps, and *Dinoderus bifoveolatus*, the genus *Lichenophanes* and *Stephanopachys substriatus* on more heavily decomposed wood, making them rather saproxylophagous species. Finally, the presence of a few species such as *Bostrychoplites cornutus, Bostrichus capucinus, Heterobostrychus brunneus, Micrapate xyloperthoides* has been confirmed on worked wood and/or manufactured wooden objects. Some species are only suspected on worked wood (*Schistoceros bimaculatus, Sinoxylon ruficorne, Stephanopachys linearis*) (Table 3.1).

M. Toriti et al., *Traces of Common Xylophagous Insects in Wood*, https://doi.org/10.1007/978-3-030-66391-9_3

Table 3.1

Species	Root	Branch	Under bark	Barkless	Trunk	Alive, declining/dying	Decomposing	Newly fallen/felled	Stored	Timber	Deciduous	Coniferous	Other	Ligneous species infested (non exhaustives)	European distribution	Cycle	Gallerie diameter
Apate monachus (Fabricius, 1775)		x	x		x	x		x	?		x		x	*Acacia, Ailanthus, Amygdalus communis, Arbutus unedo, Armeniaca vulgaris, Ceratonia siliqua, Citrus spp., Cupressus sp., Erica sp., Erythrina sp., Eucalyptus spp., Grevillea sp., Malus communis, M. domestica, Melia azedarach, Myrtus communis, Olea, Persica vulgaris, Phoenix dactylifera, Pinus pinea, Psidium guajava, Pistacia lentiscus, Pyrus spp., Prunus spp., Punica granatum, Quercus ilex, Schinus sp., Tamarix sp., Vitis*	Spain, France, Italy, Switzerland, Germany, Cyprus	?	?
Bostrychoplites cornutus (Olivier, 1790)		x	x	x	x	x		x	x	x	x		x	Exotic species ex: *Bambusoideae*	France, Germany, Italy, Spain, Sweden	?	?

Bostrichus capucinus **(Linnaeus, 1758)**	**x**	**x**	**x**	**x**	**x**	**x**		**x**	**x**	**x**	**x**	**?**		***Quercus, Juglans, Fagus, Populus, Robinia, Castanea, Corylus, Vitis, Arbutus unedo, Rosaceae, Ericaceae …*** **Maybe *Pinus***	**More common in Europe**	**1 year**	**3–5 mm**
Dinoderus bifoveolatus (Wollaston, 1858)		x	x		x	x	x	x			x		x	Exotic species ex: *Acacia*	EXOGENOUS: Austria, Belgium, Croatia, France, Germany, Poland, Slovakia, Spain, Sweden, Switzerland, The Netherlands	Around 180 days	?
Dinoderus brevis (Horn, 1878)		x	x		x	x		x	x		x		x	Exotic species; *Vitis*	EXOGENOUS: Britain, Finland, France, Germany, North European Russia, Sweden	Around 180 days	?
Dinoderus distinctus (Lesne, 1897)		x	x		x	x		x	x		x		x	Exotic species; *Vitis*	EXOGENOUS: France (Guadeloupe), Italy	Around 180 days	?
Dinoderus japonicus (Lesne, 1895)		x	x		x	x		x	x		x		x	Exotic species ex: *Arecaceae, Bambusoideae;* and *Fagus*	EXOGENOUS: Belgium, France, Germany, Italy, Sweden, The Netherlands	Around 180 days	?
Dinoderus minutus (Fabricius, 1775)		x	x		x	x		x			x		x	Exotic species: *Bambusoideae; Vitis*	EXOGENOUS: Belgium, Britain, France, Germany, Greek, Poland, Sardinia, Sicily, Slovakia, Sweden, The Netherlands	Around 180 days	?

(continued)

Table 3.1 (continued)

Species	Root	Branch	Under bark	Barkless	Trunk	Alive, declining/ dying	Decom-posing	Newly fallen/ felled	Stored	Timber	Decid-uous	Coni-ferous	Other	Ligneous species infested (non exhaustives)	European distribution	Cycle	Gallerie diameter
Dinoderus ocellaris (Stephens, 1830)		x	x		x	x		x			x		x	Exotic species: Bambusoideae; Vitis	EXOGENOUS: Belgium, Germany, France, Italy, Sweden, The Netherlands	Around 180 days	?
Ennea-desmus trispinosus (Olivier, 1795)					x	x					x		x	Arecaceae, Tamarix	France, Canaries, Croatia, Portugal, Sardinia, Sicily, Spain	?	1 mm
Heterobost-rychus aequalis (C.O. Waterhouse, 1884)	x	x	x		x	x		x	x		x	?	x	Exotic species: Acacia, Bambusoideae, Arundo spp.	Britain, France, Sweden	?	?
Heterobostr-ychus brunneus (Murray, 1867)		x	x	x	x	x		x	x	x	x		x	Exotic species: Bambusoideae, Tectona grandis; Buxus	Britain, France, Germany, Italy, Spain	1 year	?
Licheno-phanes numida (Lesne, 1899)		x	x		x		x				x	x		Eucalyptus, Pinus, Populus, Quercus spp.	Mediterranean: France, Italy, Portugal, Spain	<5 years	?
Licheno-phanes varius (Illiger, 1801)		x	x		x		x				x	x		Carpinus, Castanea, Fagus, Eucalyptus, Pinus, Populus, Quercus spp.	More common in Europe	<5 years	
Micrapate xyloper-thoides (Jacquelin du Val, 1859)		x	x		x	?		x	?		x		x	Bambusoideae, Saccharum, Vitis	France, Italy, Portugal, Sicily, Spain	?	?

Species											Host plants	Distribution		
Schistoceros bimaculatus (Olivier, 1790)	x	x	?	x	x	x	?	x	x	x	*Vitis, Tamarix, Acacia, Annona cherimola, Delonix regia, Ficus carica, Lycium, Olea europaea, Rosaceae (Malus, Prunus, Cerasus …)*	Mediterranean basin: Balearic islands, Bosnia, Croatia, Cyprus, France, Greece, Italy, Crete, Portugal, Spain	?	?
Scobicia chevrieri (Villa and Villa, 1835)	x		x	x	x		x	x	x	x	*Acacia, Amygdalus communis, Arundo, Ceratonia siliqua, Cercis, Citrus, Eucalyptus, Ficus, Hibiscus, Laurus nobilis, Morus alba, Olea, Pinus halepensis, Pistacia lentiscus, Prunus spp., Punica granatum, Quercus spp. Q. suber, Rhamnus alaternus, Ulmus, Vitis*	Mediterranean basin: Balearic islands, Croatia, Cyprus, France, Greece, Italy, Crete, Portugal, Spain	?	?
Scobicia pustulata (Fabricius, 1801)	x		x	x	x		x	x	x	x	idem *S. chevrieri*	Mediterranean basin: Balearic islands, Croatia, Cyprus, France, Greece, Italy, Crete, Portugal, Spain	?	?

(continued)

Table 3.1 (continued)

Species	Root	Branch	Under bark	Barkless	Trunk	Alive, declining/ dying	Decom- posing	Newly fallen/ felled	Stored	Timber	Decid- uous	Coni- ferous	Other	Ligneous species infested (non exhaustives)	European distribution	Cycle	Gallerie diameter
Sinoxylon conigerum (Gerstacker, 1855)											x		x	*Anacardiaceae, Combretaceae, Dipterocarpaceae, Euphorbiaceae, Lamiaceae, Lauraceae, Leguminosae, Mimosaceae, Myrtaceae, Rubiaceae, Tiliaceae, Ulmaceae*	EXOGENOUS: France, Germany, Great Britain, Italy, Poland, Russia, Spain, Ukraine	?	?
Sinoxylon perforans (Schrank, 1798)		x	?			x		x	x		x			*Quercus ilex, Q. suber*	More common in Europe	?	?
Sinoxylon ruficorne (Fahaeus, 1872)		x	?	x	x			x	x	?	x			*Acacia, Casuarina, Eucalyptus, Leguminosae*	France, Italy, Portugal	?	?
Sinoxylon sexdentatum (Olivier, 1790)		x				x		x	x		x			*Quercus, Ficus carica …*	Balearic islands, Belgium, Bosnia and Herzegovina, Canary, Croatia, France, Greece, Hungary, Italy, Crete, Portugal, Sardinia, Sicily, Slovenia, Spain, The Netherlands, Ukraine	?	?
Stephano- pachys linearis (Kugelann, 1792)		x	x			x		x	x	?	x	x		*Abies, Larix, Picea, Pinus sylvestris, Quercus*	Armenia, Austria, Azerbaijan, Belarus, Germany, Denmark, Estonia, Finland, France, Georgia, Italy, Latvia, Lithuania, Norway, Poland, Czech Republic, European Russia, Slovenia, Sweden, Switzerland, Ukraine	1–2 years	?

Species	Host plants	Distribution	Generation time	?
Stephanopachys quadricollis (Marseul, 1878)	*Abies, Larix, Picea, Pinus*	Southern and Mediterranean Europe including Croatia, France, Greece, Italy, Portugal, Spain (including Balearic Islands), Sweden (introduction), Syria, Turkey, Ukraine.	1–2 years	?
Stephanopachys substriatus (Paykull, 1800)	*Larix, Pinus cembra, Pinus sylvestris, Picea*	Albania, Austria, Belarus, Bosnia and Herzegovina, Bulgaria, Croatia, Estonia, Finland, France, Germany, Greece, the United Kingdom, Czech Republic, Hungary, Italy, Latvia, Lithuania, Macedonia, Moldova, Norway, Poland, Romania, Russia of Europe, Serbia, Montenegro, Slovakia, Slovenia, Sweden, Switzerland, Syria, Ukraine	1–2 years	?
Xylopertha praeusta (German, 1817)	*Acacia, Ficus, Eucalyptus, Pistacia, Quercus*	Mediterranean: Balearic islands, Croatia, France, Italy, Portugal, Spain	?	?
Xylopertha retusa (Olivier, 1790)	*Acer, Castanea, Ficus, Quercus, Vitis*	More common in Europe	?	?
Xyloperthella picea (Olivier, 1791)	*Acacia, Adansonia, Albizia, Bambusoideae, Hevea, Quercus, Terminalia, Ziziphus*	France, Germany, Britain, Italy, Malta, Portugal, Spain, The Netherlands	?	?

Bostrichus capucinus **(Linnaeus, 1758)**

syn.: *Bostrychus capucinus*
Capuchin Beetle

Description

The average size of adults is between 9 and 14 mm, with extreme values of 5 and 16 mm. It owes its name "capuchin" to the hood-shaped (spherical or globular) pronotum that covers its head. It has short antennae composed of 11 articles ending in a club composed of three super-imposed articles.

The elytra end in a declivity and are armed with spines (or lateral apical denticules), which makes them look more like bark beetles.

Its mandibles are very powerful. They can pierce lead plates in particular. Its elytra are brick-red while the rest of the body is black.

The white to yellowish, sometimes grey arched larva is 8–16 mm long depending on the phase of its development cycle. Its body is covered with golden and rigid hairs (bristles) on some of its segments.

Distribution

Central Asia (Russia, Kazakhstan, Mongolia, China …)—North Africa—the Mediterranean basin—northern Europe: from Scandinavia to the United Kingdom.

Bioecology

Adults are generally found from April to July, when outdoor temperatures reach 19 °C (the optimum temperature is around 25 °C). They are mainly nocturnal, and sometimes swarm in very large numbers. Both sexes bore a circular breeding chamber in the wood after mating, the female looks for wood rich in starch and other sugars such as dextrose and maltose, which she tastes beforehand.

The female deposits and distributes 50–500 eggs (150 on average) on the wood surface, in the chamber, in the wood crevices or in the xylem pores, and occasionally in old galleries.

This species ingests cellulose and hemicelluloses, preferentially starch and other sugars such as dextrose and maltose mainly concentrated in the peripheral layers of the tree (sapwood). Its food sources are close to those of Lyctinae.

The larvae develop in several larval phases, including a quiescence phase essential for the onset of pupation the following spring.

Pupation takes place below the surface of the wood, in the sapwood, and lasts 7–15 days.

Cycle of Development

A complete cycle takes about 1 year and depends on various factors: the starch content of the wood, and an average ambient temperature of 25 °C at 65% humidity, provided that the wood contains enough starch, the average ambient temperature is 25 °C and the humidity rate is 65%.

Infested Woods

The typical habitat of this species is dry open forest, with scrub and wooded parkland with felled wood lying on the ground. *B. capucinus* is also an occasional pest of stored, worked wood and furniture.

It only infests deciduous trees, freshly felled or sawn, stored with or without their bark, or even processed wood. It is most often found on felled oak (*Quercus ilex, Q. robur, Q. toza* …) and walnut (*Juglans*) trunks. However, it can also be found in many other deciduous trees and shrubs such as pine (*Pinus*), beech (*Fagus*), poplar (*Populus*), robinia (*Robinia*), chestnut (*Castanea*), vine stocks (*Vitis*), heather roots (*Erica*), some fruit trees (*Prunus* spp.) and in timber.

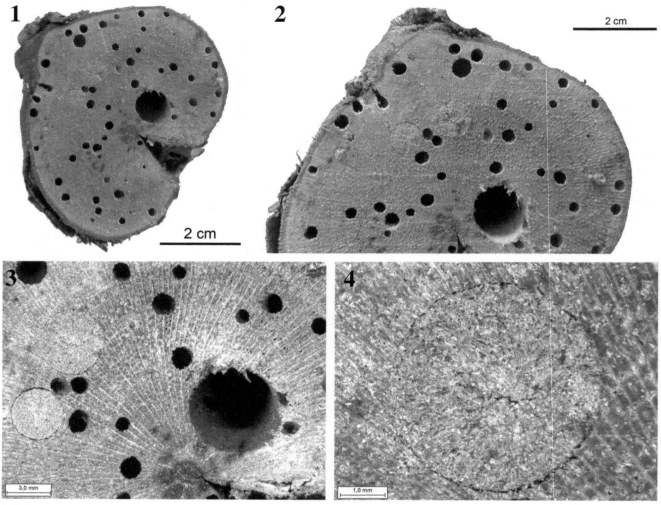

1–2. Galleries of *B. capucinus* on a vine (*Vitis vinifera*) branch (Italy, A. Crivellaro)
3. Exit holes, detail
4. Larval gallery filled with frass, detail

Damage caused by *B. capucinus* can only be detected after the first generation has emerged or slightly earlier if a swelling of the wood surface is seen, along with small piles of frass pushed away from cracks and crevices in the wood.

The galleries are round and very regular, varying from 3 to 5 mm in diameter. They generally run parallel to the direction of the grain of the wood. These galleries are larger than those of Ptinidae and Lyctidae, except those of *Xestobium rufovillosum*. The distinction is then made based on frass.

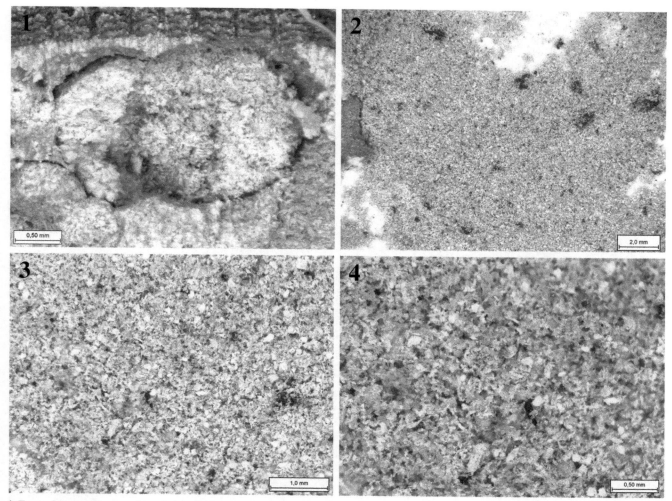

1. Frass of *B. capucinus* can be seen in galleries
2. Powder-like frass
3–4. Frass composition, detail

The frass is homogeneous, powdery, and can be highly compacted. It is fine, of the fine-flour type and therefore does not occur as a distinct pellet. It is composed of a multitude of very fine wood fibre scraps.

B. capucinus can be mistaken for members of the Lyctidae family. Differences are at the gallery level. The galleries of *B. capucinus* are bigger and do not seem to be bored in any particular wood cell, except for the sapwood area which is privileged.

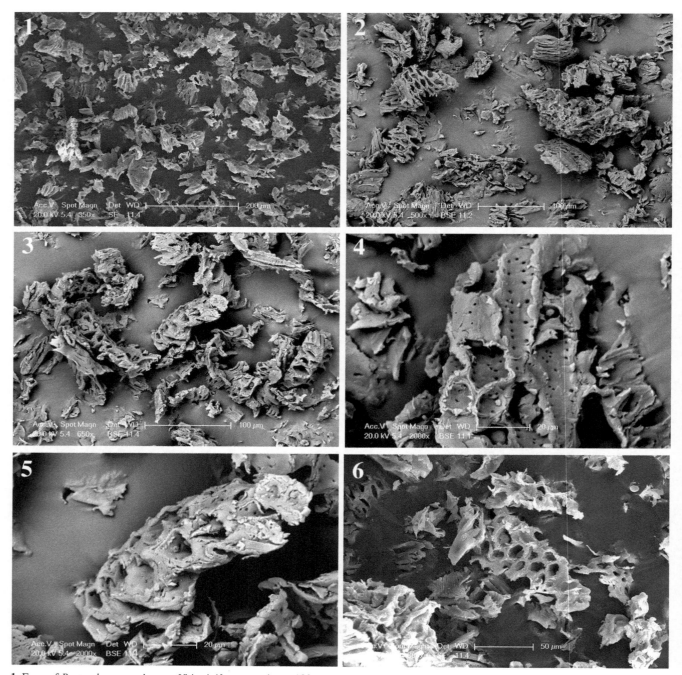

1. Frass of *Bostrychus capucinus* on *Vitis vinifera*, overview, ×100
2. Frass composition, detail: various more or less digested wood chips, ×270
3. Agglomerated digested wood fibres, detail. Quite friable, ×300
4. Wood chip composing the frass, detail: undigested ligneous vessel wastes, ×600
5. Other ligneous vessel elements, ×650
6. Scalariform perforation of *Vitis vinifera*, undigested by *B. capucinus*, ×1700

3.1.2   Cerambycidae

Presentation of the Family

Capricorns beetle (or longhorns beetle) are among the most dangerous species in the world and are important wood decomposers as they adapt to many different types of wood and contexts: carpentry and timber (*Hylotrupes bajulus*), agricultural products, stored wood (logs), various forest ecosystems, forest plantations (*Monochamus* and *Tetropium* on conifers), orchards, nurseries. In addition, invasions of exotic cerambycids such as *Anoplophora glabripennis*, *Anoplophora chinensis* and more recently *Xylotrechus chinensis* and *Callidiellum rufipenne* are increasingly worrying for native capricorn ecosystems and species because they can easily adapt, in Europe but also on the American continent. They can sometimes be considered in the same way as the American termite, which competes with the European termite, or even as the Asian hornet for bees. However, some species tend to disappear mainly because of the destruction of forest areas. Some emblematic species are officially protected, such as *Cerambyx cerdo* (strictly protected in France and Europe) or *Rosalia alpina*.

This large family includes no less than 35,000 species, divided into 4000 genera and ten subfamilies, most of which are cambio-xylemophagous, or saproxylophagous, the others are rather phytophagous or proliferate in herbaceous plants (bamboo, cactus, succulents …) (*Agapanthia*, *Phytoecia*) or in soils (*Dorcadion*). The adult individual has an elongated body and can display a wide variety of patterns and colours. The name Capricorn comes from the long, segmented antennae that often exceed the total body length. Its mandibles are robust. Their size varies from a few millimetres to several centimetres depending on the species (generally 8–50 mm).

Larval life is much longer than adult life. This is why it is so essential to have data about the damage caused by these species to woods. The larva is subcylindrical, slightly flattened and sclerotized. Its length varies from 5 to 220 mm depending on the species. It is generally white with an orange to black sclerotized head.

Pupation usually occurs in the plant species where larval life took place. Adults can be observed in the spring or summer on log piles, logs, flowers or shrubs. Females lay eggs in damaged wood, in bark cracks or in the soil around roots, in fallen wood. The larvae emerge after a few weeks and feed on the host plant. The larval-stage to adult-stage cycle generally lasts about 1 year for small species and 3–4 years for larger species. If environmental conditions are not met, the cycle can last up to 6 years (*Hylotrupes bajulus*). The galleries of capricorns are a discriminating criterion because they are the only ones to be oval or elliptical. Thus, some capricorns are considered as primary pests (*Saperda*), i.e., they infest living wood. The larvae of the majority of species bore sub-cortical galleries (*Rhagium*, *Acanthocinus*) and bore into the sapwood (*Clytus*, *Callidium*), sometimes as far as the heartwood (*Cerambyx*, *Monochamus*), which makes them parasites of weakened or freshly felled trees (secondary pests). Some of them live on barkless trees, and weaken stored wood and carpentry wood (*Hylotrupes bajulus*, *Arhopalus*, *Clytus pilosus*). Other species prefer dead wood in more or less advanced decomposition (*Ergates faber*, *Leptura aurulenta*, *Oxymirus cursor*, *Rosalia alpina*). On the whole, adults live in and feed on flowers.

The European fauna includes about 700 species. Diversity is greater in warmer regions. About 260 species are currently listed in France, 275 in Italy, 220 in Croatia, about 200 in Germany, 75 in Denmark and 70 in the United Kingdom. As our study began in France, we carried out an inventory of strictly xylophagous species among this corpus made up of 260 species, based on several works such as Berger (2012). Strictly phytophagous species and flower-dwellers have been withdrawn from the presentation of this family, as well as certain exogenous species recently introduced or proliferating in Europe. For further information, it should be noted that much research on this family has already been carried out, e.g., Picard (1929), Butovitsch (1939), Duffy (1953), Linsley (1959), Demelt (1966), Villiers (1978) …

The following tables list the data collected to date about the typical biology of each species: preferred wood species, infested part(s) of the wood, general state of the wood (alive, dead, stored, worked …), length of the development cycle, average size of the galleries …

Among the species of this corpus are 112 species living exclusively on deciduous trees (e.g., *Aegomorphus*, *Aegosoma*, *Akimerus*, *Albana*, *Anaglyptus*, *Anoplophora*, *Anoplodera*, *Aromia*, *Callimus*, *Cerambyx*, *Chlorophorus*, *Clytus*, *Deilus*, *Deroplia*, *Dinoptera*, *Grammoptera*, *Menesia*, *Oplosia*, *Poecilium*, *Pyrrhidium*, *Saperda*, *Tetrops*, *Trichoferus*, *Xylotrechus* …), 45 on coniferous trees (e.g., *Acanthocinus*, *Acmaeops*, *Arhopalus*, *Callidium*, *Ergates*, *Icosium*, *Molorchus*, *Monochamus*, *Obrium*, *Pachyta*, *Pogonocherus*, *Semanotus*, *Tetropium*, *Tragosoma* …), and 36 species that can be found on both deciduous and coniferous trees in spite of an often notable preference for one of the two families (e.g., *Alosterna*, *Callidium violaceum*, *Evodinus*, *Gaurotes*, *Hylotrupes bajulus*, *Leiopus nebulosus*, *Leptura*, *Mesosa*, *Morimus*, *Parmena*, *Penichroa*, *Rhagium*, *Ropalopus*, *Stenurella* …).

The biology of some species, many of which are rare, is still poorly known or uncertain. Other species would deserve a regular and wide-ranging check-up of their distribution across the different regions of Europe because they easily adapt to the plant species they infest. This work is in progress, and carried out on recent and current periods by entomologists specialised in cerambycids.

We divided Cerambycidae into four types:

- Cerambycidae type 1: primary pest, only cambiophagous
- Cerambycidae type 2: secondary pest, xylemophagous
- Cerambycidae type 3: from cambium to heartwood, sometimes decaying wood
- Cerambycidae type 4: on dry and/or worked woods

Caution is required when identifying the galleries and vermilions of this family. Sometimes it is better to stay within the sub-family. It is possible to differentiate between fresh-wood and dry-wood capricorn beetles. And within fresh wood Capricorn beetles, it is possible to tell if it is a strict cambiophagous one (primary pest) or cambio-xylemophagous one (secondary pest) (Table 3.2).

Table 3.2

Species	Subfamily	Tribe	Root	Branch	Under bark	Bark-less	Trunk	Alive, declining/ dying	Decom-posing	Newly fallen/ felled	Stored	Timber	Decid-uous	Conifer-ous	Ligneous species infested (non exhaustive list)	European distribution	Cycle	Gallerie diameter
Acanthocinus aedilis (Linnaeus, 1758)	Lamiinae	Acanthocinini		x	x		x			x				x	*Pinus* (often), *Abies, Picea, Larix, Cedrus*	All Europe except Ireland, Mediterranean islands	2 years	>1 cm
Acanthocinus griseus (Fabricius, 1792)	Lamiinae	Acanthocinini		x	x		x			x				x	*Pinus sylvestris, P. maritima, P. halepensis, Picea, Abies*	All Europe except Ireland, the UK, Portugal, Belgium, The Netherlands	1 year	>1 cm
Acanthocinus reticulatus (Razoumowsky, 1789)	Lamiinae	Acanthocinini		x	x		x			x				x	Mainly: *Abies* More rarely: *Pinus, Picea*	Austria, Bosnia and Herzegovina, Bulgaria, Croatia, Czech Republic, France, Germany, Greece, Italy, Poland, Romania, Slovakia, Slovenia, Spain, Switzerland, Ukraine	1 year	>1 cm
Acmaeops marginattus (Fabricius, 1781)	Lepturinae	Rhagiini		x	x		x	x		x	x			x	*Pinus, Picea, Abies*	Austria, Belgium, Bosnia and Herzegovina, Czech Republic, Estonia, Finland, France, Germany, Greece, Italy, Lithuania, Poland, Slovakia, Spain, Sweden, Switzerland, The Netherlands, Ukraine	2 years	>0.5 cm
Acmaeops pratensis (Laicharting, 1784)	Lepturinae	Rhagiini			x		x			x	x			x	*Pinus, Picea, Abies*	All Europe except Croatia, Denmark, Ireland, the UK, Portugal, Belgium, The Netherlands	2 years	>0.5 cm
Acmaeops septentrionis (Thomson, 1866)	Lepturinae	Rhagiini		x	x		x	x		x	x			x	*Pinus, Picea*	Austria, Bulgaria, Czech Republic, Estonia, Finland, France, Germany, Italy, Latvia, Liechtenstein, Lithuania, Norway, Poland, Romania, Slovakia, Sweden, Switzerland, Ukraine	2 years	>0.5 cm

(continued)

Table 3.2 (continued)

Species	Subfamily	Tribe	Root	Branch	Under bark	Bark-less	Trunk	Alive, declining/dying	Decom-posing	Newly fallen/felled	Stored	Timber	Decid-uous	Conifer-ous	Ligneous species infested (non exhaustive list)	European distribution	Cycle	Gallerie diameter
Acmaeops smaragdulus (Fabricius, 1792)	Lepturinae	Rhagiini			x		x	x	x	x	x			x	Pinus, Picea, Abies, Larix	Finland, France, Latvia, Lithuania, Norway, Poland, Sweden, Switzerland	2 years	>0.5 cm
Aegomorphus clavipes (Schrank, 1781)	Lamiinae	Acanthoderini		x	x		x			x			x		Mainly: Populus, Betula, Juglans, Fagus More rarely: Prunus, Pinus, Alnus, Acer, Morus, Corylus, Salix, Tilia, Quercus, Ailanthus …	All Europe except Ireland, the UK, Portugal, Belgium, The Netherlands	2 years	>0.5 cm
Aegomorphus francottei (Sama, 1994)	Lamiinae	Acanthoderini			x		x		x	x	x			x	Quercus, Castanea	Balkany region, France, Greece	2 years	>0.5 cm
Aegosoma scabricorne (Scopoli, 1763)	Prioninae	Aegosomatini			x		x	x						x	Mainly: Fagus More rarely: Tilia, Quercus, Acer, Fraxinus, Ulmus, Salix, Populus …	Albania, Austria, Bosnia and Herzegovina, Bulgaria, Croatia, Czech Republic, France, Germany, Greece, Hungary, Italy, Crete, Macedonia, Moldova, Romania, Sardinia, Sicily, Slovakia, Slovenia, Spain, Switzerland, Ukraine	2–3 years	Around 2 cm
Akimerus schaefferi (Laicharting, 1784)	Lepturinae	Rhagiini	x		x		x	x		x			x		Ulmus, Quercus, Fagus, Fraxinus, Acer, Malus, Prunus	Austria, Bulgaria, Central European Russia, Croatia, Czech Republic, France, Germany, Greece, Hungary, Moldova, Poland, Portugal, Slovakia, Slovenia, Spain, Switzerland, Ukraine	2–3 years	>1 cm
Albana m-griseum (Mulsant, 1846)	Lamiinae	Pteropliini		x			x			x			x		Genista scorpius, G. cinerea, Spartium junceum, Sarothamnus scoparius	France, Spain	1 year	>0.5 cm

Species	Subfamily	Tribe									Host plants	Distribution	Generation	Size
Alosterna tabacicolor (De Geer, 1775)	Lepturinae	Lepturini	x	x	x	x	x	x	x	x	Diverses	All Western Europe except Portugal, Sardinia, Sicily	2 years	>0.5 cm
Anaesthetis testacea (Fabricius, 1781)	Lamiinae	Desmiphorini	x	x	x	x	x	x	x	x	Mainly: *Castanea, Juglans, Quercus* More rarely: *Alnus, Salix, Betula, Populus, Corylus, Prunus, Malus, Pyrus, Rosa* …	All Western Europe except Ireland, the UK, Denmark	2 years	>0.5 cm
Anaglyptus gibbosus (Fabricius, 1787)	Cerambycinae	Clytini	x		x		x			x	*Quercus petraea, Q. ilex, Acer, Fagus, Sorbus, Ficus, Crataegus* …	Croatia, France, Greece, Italy, Sicily, Slovenia, Spain, Switzerland	2 years	Around 0.5 cm
Anaglyptus mysticus (Linneaus, 1758)	Cerambycinae	Clytini	x	x	x	x	x	x	x	x	*Quercus, Acer, Alnus, Ulmus, Fagus, Carpinus, Tilia, Prunus, Pyrus, Ficus, Juglans* …	All Western Europe except Ireland, Mediterranean Islands, Portugal	2 years	Around 0.5 cm
Anastrangalia dubia (Scopoli, 1763)	Lepturinae	Lepturini	x	x	x	x	x	x		x	*Picea, Pinus maritima, P. sylvestris, Abies, Larix*	All Western Europe except Denmark, Ireland, Luxembourg, the Mediterranean Islands, Portugal, The Netherlands, the UK	2 years	>1 cm
Anastrangalia reyi (Heyden, 1889)	Lepturinae	Lepturini	x	x	x	x	x	x	x	x	*Picea, Pinus maritima, P. sylvestris, Abies, Larix*	All Western Europe except Balkany region, Belgium, Denmark, Ireland, Luxembourg, the Mediterranean islands, Portugal, The Netherlands, the UK	2 years	>1 cm
Anastrangalia sanguinolenta (Linneaus, 1761)	Lepturinae	Lepturini	x	?	x		x			x	*Pinus, Abies, Picea*	All Western Europe except Belgium, Denmark, Ireland, the Mediterranean Islands, Portugal	2 years	>0.5 cm

(continued)

Table 3.2 (continued)

Species	Subfamily	Tribe	Root	Branch	Under bark	Bark-less	Trunk	Alive, declining/dying	Decomposing	Newly fallen/felled	Stored	Timber	Deciduous	Coniferous	Ligneous species infested (non exhaustive list)	European distribution	Cycle	Gallerie diameter
Anisorus quercus (Götz, 1783)	Lepturinae	Rhagiini	x		x			x		x			x		*Quercus, Acer*	Austria, Belarus, Belgium, Bosnia and Herzegovina, Bulgaria, Croatia, Czech Republic, France, Germany, Greece, Hungary, Italy, Luxembourg, Macedonia, Moldova, Poland, Romania, Slovakia, Spain, Switzerland, The Netherlands, Ukraine	2–3 years	>1 cm
Anoplophora chinensis (Forster, 1771)	Lepturinae	Lepturini	x	x	x		x	x		x	x		x		Softwood deciduous trees: *Acer, Alnus, Betula, Citrus, Fagus, Populus, Prunus, Salix, Ulmus* …	EXOGENOUS	2 years	>1 cm
Anoplophora glabripennis (Motschulsky, 1854)	Lepturinae	Lepturini	x	x	x		x	x		x	x		x		Softwood deciduous trees: *Acer, Alnus, Betula, Fagus, Populus, Salix, Ulmus* …	EXOGENOUS	2 years	>1 cm
Anoplodera rufipes (Schaller, 1783)	Lepturinae	Lepturini		x	x			x	x		x		x		*Fagus, Quercus, Betula*	All Western Europe except Denmark, Ireland, Luxembourg, Portugal, The Netherlands, the UK and Mediterranean islands	2 years	>0.5 cm
Anoplodera sexguttata (Fabricius, 1775)	Lepturinae	Lepturini		x	x				x		x		x		*Quercus, Alnus, Fagus, Carpinus*	All Western Europe except Portugal and the Mediterranean islands	2 years	>0.5 cm
Arhopalus ferus (Mulsant, 1839)	Aseminae	Asemini	x				x				x	x		x	Mainly: *Pinus* More rarely: *Abies, Picea, Larix*	All Western Europe except Albania, Ireland, Switzerland	1 year minimum	1–1.5 cm
Arhopalus rusticus (Linnaeus, 1758)	Aseminae	Asemini	x				x				x	?		x	*Pinus, Abies, Larix*	All Western Europe except Albania, Ireland, the Balearic islands, Crete and Sardinia	2–3 years	1–1.5 cm

Species	Subfamily	Tribe									Host plants	Distribution	Development	Size
Arhopalus syriacus (Reitter, 1895)	Aseminae	Asemini	x				x		x	?	*Pinus pinaster, P. laricio, P. salzmannii, P. halepensis, P. pinea*	Mediterranean basin: Azores Is …, Balearic Is., Canary Is., Croatia, France, Greece, Italy, Crete, Portugal, Sardinia, Sicily, Spain	2 years	Around 1 cm
Aromia bungii (Faldermann, 1835)	Cerambycinae	Callichromatini	x	x	x		x	x	x	x	*Rosaceae* Mainly: *Prunus* spp.	EXOGENOUS: France, Germany, Spain?	2–4 years	>1 cm
Aromia moschata (Linnaeus, 1758)	Cerambycinae	Callichromatini	x	x	x		x		x	x	Mainly: *Salix* More rarely: *Alnus, Populus, Acer*	All Western Europe except Albania, Balearic Is. and Sardinia	3 years minimum	Around 2 cm
Asemum striatum (Linnaeus, 1758)	Aseminae	Asemini	x		x	x	x	x		x	*Pinus*	All Western Europe except Albania, Ireland, Portugal and Mediterranean islands	1–2 years	>1 cm
Callidiellum rufipenne (Motschulsky, 1861)	Cerambycinae	Callidiini	x	x	x		x	x		x	*Cupressus, Thuya, Juniperus*	EXOGENOUS: France, Italy, Spain	1–2 years	>1 cm
Callidium aeneum (De Geer, 1775)	Cerambycinae	Callidiini	x	x	x		x	x		x	*Abies, Picea, Pinus, Larix*	Balkany region, Austria, Belgium, Czech Republic, France, Germany, Greece, Italy, Luxembourg, Switzerland, The Netherlands …	2 years	Around 1 cm
Callidium coriaceum (Paykull, 1800)	**Cerambycinae**	**Callidiini**	x	x	x		x			x	**Mainly: *Picea* More rarely: *Abies, Larix, Pinus***	**Austria, Belarus, Bosnia and Herzegovina, Czech Republic, Estonia, Finland, France, Germany, Greece, Hungary, Italy, Latvia, Liechtenstein, Lithuania, Luxembourg, Norway, Poland, Romania, Slovakia, Slovenia, Sweden, Switzerland, Ukraine**	**1–2 years**	**Around 1 cm**

(continued)

Table 3.2 (continued)

Species	Subfamily	Tribe	Root	Branch	Under bark	Bark-less	Trunk	Alive, declining/ dying	Decom- posing	Newly fallen/ felled	Stored	Timber	Decid- uous	Conifer- ous	Ligneous species infested (non exhaustive list)	European distribution	Cycle	Gallerie diameter
Callidium violaceum (Linnaeus, 1758)	Ceramby- cinae	Callidiini		x	x		x			x	x	x	x	x	Mainly: *Picea, Pinus, Larix* More rarely: *Alnus, Fagus, Quercus, Salix, fruitiers …*	All Western Europe except Greece, Portugal, Spain, Mediterranean islands	2 years	Around 1 cm
Callimus abdominalis (Olivier, 1795)	Ceramby- cinae	Stenopterini		x	x		x			x			x		*Acer, Q. ilex, Q. suber; Tilia, Fagus, Prunus, Corylus, Pistacia, Rosa* …	Croatia, France, Greece, Sardinia, Sicily, Spain	2 years	>0.5 cm
Callimus angulatus (Schrank, 1789)	Ceramby- cinae	Stenopterini		x	x		x			x			x		*Quercus, Q. ilex, Fraxinus, Ficus, Castanea, Ostrya* …	Austria, Balkany region, Czech Republic, France, Germany, Greece, Hungary, Italy, Crete, Poland, Sicily, Spain, Switzerland, Ukraine	2 years	>0.5 cm
Cerambyx cerdo (Linnaeus, 1758)	**Ceramby- cinae**	**Cerambycini**		x	x		x			x	x	?	x		*Quercus*	**All Western Europe except Denmark, Ireland, Crete**	**3 years**	**<3 cm**
Cerambyx miles (Bonelli, 1823)	Ceramby- cinae	Cerambycini		x	?		x			x	x		x		*Quercus ilex* More rarely: other *Quercus, Rosaceae, Vitis*	Balkany region except Albania; France, Greece, Italy, Portugal, Spain, Switzerland	2–3 years?	Around 2 cm
Cerambyx scopolii (Fuesslins, 1775)	Ceramby- cinae	Cerambycini		x	?		?	x		x	?		x		*Rosaceae*	All Western Europe except Ireland, the UK	2–3 years	1–1.5 cm
Cerambyx welensii (Küster, 1846)	Ceramby- cinae	Cerambycini		x			x			x	x		x		*Quercus ilex* and other *Quercus*	Balkany region except Albania; France, Greece, Italy, Crete, Portugal, Sicily, Spain	2–3 years	2.5–3 cm

Species	Subfamily	Tribe										Host plants	Distribution	Generation	Size
Chlorophorus glabroma-culatus (Goeze, 1777)	Ceramby-cinae	Clytini	x			x			x	x	x	*Quercus, Castanea, Robinia, Acacia dealbata, Vitis, Acer, Tilia, Ulmus, Alnus, Populus, Salix, Prunus, pyrus, Ficus, Juglans …*	Belgium, Croatia, France, Greece, Italy, Malta, Sardinia, Sicily, Slovenia, Spain, Switzerland	2 years	>1 cm
Chlorophorus glaucus (Fabricius, 1781)	Ceramby-cinae	Clytini	x			x			x	?	x	Mainly: *Acacia dealbata* More rarely: *Quercus ilex; Ulmus, Celtis, Salix, Malus*	Balearic Is., France, Portugal, Sardinia, Spain	2 years	>1 cm
Chlorophorus figuratus (Scopoli, 1763)	Ceramby-cinae	Clytini	x			x			x	x	x	*Diverses*	All Western Europe except Albania, Balearic Is., Denmark, Ireland, Crete, Portugal, Sicily, The Netherlands, the UK	2 years	>1 cm
Chlorophorus herbstii (Brahm, 1790)	Ceramby-cinae	Clytini	x		x	x					x	Mainly: *Tilia* More rarely: *Quercus, Ulmus, Castanea, Betula, Carpinus, Crataegus, Corylus*	All Western Europe except Albania, Denmark, Ireland, Italy, Luxembourg, Macedonia, Mediterranean islands, Portugal, Sicily, The Netherlands, the UK	2 years	>1 cm
Chlorophorus ruficornis (Olivier, 1790)	Ceramby-cinae	Clytini	x			x			x	x	x	*Quercus ilex*	France, Portugal, Spain	2 years	>1 cm
Chlorophorus sartor (Müller, 1766)	Ceramby-cinae	Clytini	x		x	x			x	x	x	*Quercus, Castanea, Robinia …*	All Western Europe except Balearics Is., Denmark, Ireland, The Netherlands, the UK	2 years	Around 0.5 cm
Chlorophorus trifasciatus (Fabricius, 1781)	Ceramby-cinae	Clytini	x			x			x	x	x	*Quercus, Fabaceae (Ononix, Dorycnium)*	Austria, Balearic Is., Bulgaria, Croatia, France, Greece, Hungary, Italy, Portugal, Romania, Sardinia, Sicily, Slovenia, Spain, Switzerland	2 years	Around 0.5 cm

(continued)

Table 3.2 (continued)

Species	Subfamily	Tribe	Root	Branch	Under bark	Bark-less	Trunk	Alive, declining/ dying	Decom-posing	Newly fallen/ felled	Stored	Timber	Decid-uous	Conifer-ous	Ligneous species infested (non exhaustive list)	European distribution	Cycle	Gallerie diameter
Chlorophorus varius (Müller, 1766)	Ceramby-cinae	Clytini		x			x				x	x	x		Ficus, Pistacia, Robinia, Acacia dealbata, Quercus, Juglans, Castanea, Vitis, Genista …	All Western Europe except Balearics Is., Denmark, Ireland, Crete, Luxembourg, Portugal, The Netherlands, the UK	2 years	Around 0.5 cm
Clytus arietis (Linnaeus, 1758)	Ceramby-cinae	Clytini		x			x	x		x	x		x		Diverses	All Western Europe except Albania, Crete, Sardinia, Sicily	1 year?	>1 cm
Clytus lama (Mulsant, 1847)	Ceramby-cinae	Clytini			x		x			x	x			x	Picea, Abies, Larix	Austria, Bosnia and Herzegovina, Bulgaria, Croatia, Czech Republic, France, Germany, Greece, Hungary, Italy, Liechtenstein, Poland, Romania, Slovakia, Slovenia, Switzerland, Ukraine	1 year?	>1 cm
Clytus rhamni (Germar, 1817)	Ceramby-cinae	Clytini		x				x		x			x		Prunus, Quercus, Robinia …	All Western Europe except Balearics Is., Denmark, Ireland, Crete, Luxembourg, The Netherlands, the UK	1 year?	>1 cm
Clytus tropicus (Panzer, 1795)	Ceramby-cinae	Clytini		x	x	x	x	x	x	x	x		x		Mainly: Quercus More rarely: Pyrus, Prunus, Cerasus	Austria, Belgium, Bulgaria, Croatia, Czech Republic, France, Germany, Greece, Hungary, Macedonia, Moldova, Poland, Romania, Slovakia, Spain, Switzerland, Ukraine	2 years	Around 1 cm
Cornumutila quadrivittata (Gebler, 1830)	Lepturinae	Lepturini				x	x	x						x	Picea, Pinus, Larix, Abies	Austria, Czech Republic, France, Italy, Poland, Romania, Slovakia, Ukraine, Switzerland	3 years mini-mum	>0.5 cm

Species	Subfamily	Tribe									Host plants	Distribution	Generation	Size
Cyrtoclytus capra (Germar, 1824)	Ceramby-cinae	Clytini	x		x	x	x	x		x	Mainly: *Alnus, Salix caprea* More rarely: *Quercus, Betula, Ulmus, Juglans, Acer*	Austria, Bosnia and Herzegovina, Czech Republic, Estonia, France, Germany, Greece, Latvia, Lithuania, Moldova, Poland, Romania, Slovakia, Switzerland, Ukraine	2 years	>1 cm
Deilus fugax (Olivier, 1790)	Ceramby-cinae	Deilini	x		x	x	x			x	*Genista, Cytisus* Grasses: *Spartium, Sarothamnus Calycotome*	Austria, Belgium, Bosnia and Herzegovina, Bulgaria, Croatia, Cyprus, Czech Republic, France, Greece, Hungary, Italy, Crete, Moldova, Poland, Portugal, Romania, Sardinia, Sicily, Slovakia, Slovenia, Spain, Switzerland, Ukraine	2 years	>0.5 cm
Deroplia genei (Aragona, 1830)	Lamiinae	Desmiphorini	x		x		x				*Quercus*	Austria, Balkany region except Macedonia; Cyprus, Czech Republic, France, Germany, Hungary, Italy, Sicily, Spain, Ukraine	2 years	>0.5 cm
Deroplia troberti (Mulsant, 1843)	Lamiinae	Desmiphorini	x		x		x				*Quercus ilex, Q. suber; Nerium oleander*	Bosnia and Herzegovina, Croatia, France, Greece, Italy, Mediterranean Is., Portugal, Spain	1 year?	>0.5 cm
Dinoptera collaris (Linnaeus, 1758)	Lepturinae	Rhagiini	x	x	x		x	x			Mainly: *Castanea, Fagus, Quercus* More rarely: *Pirus, Malus, Populus, Robinia, Acer; Fraxinus, Viburnum, Rubus*	All Western Europe except Balearics is., Kriti, Sardinia	2 years	>0.5 cm

(continued)

Table 3.2 (continued)

Species	Subfamily	Tribe	Root	Branch	Under bark	Bark-less	Trunk	Alive, declining/dying	Decomposing	Newly fallen/felled	Stored	Timber	Decid-uous	Conifer-ous	Ligneous species infested (non exhaustive list)	European distribution	Cycle	Gallerie diameter
Drymochares truguii (Mulsant, 1847)	Aseminae	Saphanini		x			x		x	x	x	x	x		Mainly: *Corylus* More rarely: *Alnus, Fagus, Ostrya*	France, Italy	3 years mini-mum?	Around 1 cm
Ergates faber (Linnaeus, 1760)	Prioninae	Ergatini	x		x		x			x	x			x	Mainly: *Pinus, Picea* More rarely: *Abies, Cedrus, Larix*	Albania, Balearic Is., Bosnia and Herzegovina, Bulgaria, Croatia, Czech Republic, France, Greece, Hungary, Italy, Latvia, Lithuania, Poland, Portugal, Sicily, Slovakia, Slovenia, Spain, Sweden, Switzerland, The Netherlands, Ukraine	2 years	Around 2–3 cm
Evodinus clathratus (Fabricius, 1792)	Lepturinae	Rhagiini		x	x		x		x				x	x	*Picea, Fagus, Salix, Alnus*	Austria, Bosnia and Herzegovina, Bulgaria, Croatia, Czech Republic, France, Germany, Hungary, Italy, Liechtenstein, Moldova, Poland, Romania, Slovakia, Slovenia, Switzerland, Ukraine	2 years	>1 cm
Exocentrus adspersus (Mulsant, 1846)	Lamiinae	Pogonocherini		x	x			x		x	x		x		Mainly: *Quercus, Castanea* More rarely: *Carpinus, Corylus, Alnus, Juglans, Robinia, Crataegus*	All Western Europe except Balearics Is., Denmark, Ireland, Crete, Portugal, Sardinia, the UK	2 years	>0.5 cm
Exocentrus lusitanus (Linnaeus, 1767)	Lamiinae	Pogonocherini		x	x			x		x	x		x		*Tilia*	All Western Europe except Mediterranean is., Ireland, Luxembourg, Portugal, The Netherlands, the UK	1–2 years	>0.5 cm

Species	Subfamily	Tribe											Host plants	Distribution	Life cycle	Size
Exocentrus punctipennis (Mulsant et Guillebeau, 1856)	Lamiinae	Pogonocherini	x	x	x		x	x	x	x		x		*Ulmus (Salix)*	All Western Europe except Balearics Is., Denmark, Ireland, Crete, Luxembourg, Macedonia, Portugal, Sicily, The Netherlands, the UK	2 years	>0.5 cm
Gaurotes virginea (Linnaeus, 1758)	Lepturinae	Rhagiini	x	x	x	x	x	x	x	x	x	x	x	*Picea, Pinus, Abies, Larix …Quercus, Juglans*	France, Germany, Greece, Italy, Switzerland and Central and Eastern Europe	2 years	>1 cm
Glaphyra marmottani (Brisout, 1863)	Cerambycinae	Molorchini	x			x Brûlé							x	*Pinus salzmannii, P. Halepensis, P. sylvestris*	Austria, Czech Republic, France, Germany, Italy, Poland, Slovakia, Spain, Switzerland	1 year?	>0.5 cm
Glaphyra umbellatarum (Schreber, 1759)	Cerambycinae	Molorchini	x	x	x	x	x	x	x	x	x	x	x	*Rosaceae*	All Western Europe except Albania, Mediterranean is., Ireland, Portugal	2 years	>0.5 cm
Gracilia minuta (Fabricius, 1781)	Cerambycinae	Graciliini	x	x			x	x	x		x	x	x	Diverses	All Western Europe except Denmark, Ireland, Macedonia	1 year	>0.5 cm
Grammoptera abdominalis (Stephens, 1831)	Lepturinae	Lepturini	x	x	x	x	x	x	x	x	x	x	x	Mainly: *Quercus*	All Western Europe except Albania, Mediterranean is., Ireland, Luxembourg	2 years	>0.5 cm
Grammoptera ruficornis (Fabricius, 1781)	Lepturinae	Lepturini	x	x		x	x	x	x	x	x	x	x	Diverses	All Western Europe except Balearic Is., Sardinia, Crete	1 year	>0.5 cm
Grammoptera ustulata (Schaller, 1783)	Lepturinae	Lepturini	x	x	x		x	x	?	x	x	x	x	*Quercus, Castanea, Juglans, Acer …*	All Western Europe except Balearic Is., Ireland, Luxembourg, Sardinia, Crete	1 year	>0.5 cm
Herophila tristis (Linnaeus, 1767)	Laminae	Phrissomini	x	x	x		x			x	x	x	x	Mainly: *Ficus* More rarely: *Morus, Salix, Tamarix, Populus, Acer, Fraxinus … Cupressus*	Balkany region except Albania; Austria, France, Greece, Italy, Crete, Sardinia, Sicily, Switzerland	1 year?	>0.5 cm
Hesperophanes sericeus (Fabricius, 1787)	Cerambycinae	Hesperophanini	x		x		x	x	x	x	x	x	x	*Ficus, Quercus, Vitis, Lentisque, Olea, Juglans, Rhamnus (Paliure), Ceratonia, Rosaceae*	Balearic Is., Bosnia and Herzegovina, Croatia, France, Greece, Italy, Crete, Malta, Portugal, Sardinia, Sicily, Spain, Ukraine	2–3 years	1–1.5 cm

(continued)

Table 3.2 (continued)

Species	Subfamily	Tribe	Root	Branch	Under bark	Bark-less	Trunk	Alive, declining/dying	Decom-posing	Newly fallen/felled	Stored	Timber	Decid-uous	Conifer-ous	Ligneous species infested (non exhaustive list)	European distribution	Cycle	Gallerie diameter
Hylotrupes bajulus (Linnaeus, 1758)	Ceramby-cinae	Callidiini		x	x	x	x			x	x	x	x	x	Mainly: *Picea, Abies, Pinus* Accommodate: *Populus, Tamarix, Genista, Acacia* et bois ouvrés	All Western Europe	2 years Up to 6 years	Around 1 cm
Icosium tomentosum (Lucas, 1854)	Ceramby-cinae	Achrysonini		x	x		x			x	x			x	Cupressaceae	Balearic Is, Bosnia and Herzegovina, Croatia, Cyprus, France, Greece, Italy, Crete, Malta, Sardinia, Spain	2 years	>1 cm
Judolia sexmaculata (Linnaeus, 1758)	Lepturinae	Lepturini	x				x		x	x	x		x	x	*Pinus, Abies, Picea, Larix* … *Tilia, Populus*	Austria, France, Germany, Italy, the UK and Central and Eastern Europe	2 years	>0.5 cm
Lamia textor (Linnaeus, 1758)	Lamiinae	Lamiini	x		x		x	x					x		Mainly: *Salix, Populus* More rarely: *Betula, Morus, Alnus*	All Western Europe except Mediter-ranean is., Portugal	2–3 years	>1 cm
Leioderes kollari (Redtenbacher, 1849)	Ceramby-cinae	Callidiini		x	x		x			x	x		x		Mainly: *Acer, Carpinus, Corylus* More rarely: *Ulmus, Quercus*	Balkany region except Albania, Austria, Czech Republic, France, Germany, Greece, Italy, Sicily and Eastern Europe	1–2 years	>1 cm
Leiopus fermoratus (Fairmaire, 1859)	Lamiinae	Acanthocinini		x	x		x		x	x	x		x		*Castanea, Viscum, Carpinus, Tilia, Juglans, Ficus*	France, Italy, Sicily	1 year?	>0.5 cm
Leiopus nebulosus (Linnaeus, 1758)	Lamiinae	Acanthocinini		x	x		x		x	x	x		x	x	*Corylus, Juglans, Populus, Salix, Carpinus, Quercus, Acer, Betula, Alnus, Castanea, Fagus, Ulmus, Tilia,* Fruit trees; *Picea, Abies*	All Western Europe except Mediter-ranean is. (except Sicily), Ireland, Macedonia	1–2 years	>0.5 cm
Leiopus punctulatus (Paykull, 1800)	Lamiinae	Acanthocinini		x	x		x			x	x		x		*Populus*	Austria, Czech Republic, France, Germany, Poland, Switzerland	1–2 years	>0.5 cm

Species	Subfamily	Tribe												Host plants	Distribution	Generation time	Size
Leptura aethiops (Poda, 1761)	Lepturinae	Lepturini		x	?	x		x		x	?	x	x	*Alnus, Salix, Tilia, Corylus, Betula … Pinus*	All Western Europe except Mediterranean is., Ireland, Macedonia, Portugal, Spain, the UK	1–2 years?	>1 cm
Leptura annularis (Fabricius, 1801)	Lepturinae	Lepturini		x	x	x	?	x	x	x	?	x	x	Mainly coniferous	Austria, Belgium, Bosnia and Herzegovina, Croatia, Czech Republic, France, Germany, Hungary, Italy, Latvia, Lithuania, Poland, Romania, Slovakia, Switzerland	More 1 year	>1 cm
Leptura aurulenta (Fabricius, 1792)	Lepturinae	Lepturini	x			x		x					x	*Fagus, Alnus, Quercus, Castanea, Salix, Betula, Populus, Juglans, Ulmus, Prunus, Aesculus*	Balkany region, Austria, Belgium, Czech Republic, France, Germany, Greece, Ireland, Italy, Luxembourg, Portugal, Spain, Switzerland, The Netherlands, the UK	2 years	>1 cm
Leptura quadrifasciata (Linnaeus, 1758)	Lepturinae	Lepturini		x		x		x				x	?	*Alnus, Betula, Populus, Salix, Quercus, Fagus, Corylus, Carpinus, Ulmus, Platanus, Sorbus, Prunus*	All Western Europe except Mediterranean is., Luxembourg, Portugal	2–3 years	>1 cm
Lepturobosca virens (Linnaeus, 1758)	Lepturinae	Lepturini		x	x	x		x	x	x	x	x	x	*Pinus, Picea, Abies*	Austria, Bosnia and Herzegovina, Bulgaria, Czech Republic, France, Germany, Hungary, Italy, Latvia, Lithuania, Poland, Romania, Slovakia, Slovenia, Spain, Switzerland, Ukraine	More 1 year	>1 cm
Menesia bipunctata (Zoubkoff, 1829)	Lamiinae	Saperdini		x	x	x		x	x			x		*Frangula*	Austria, Bosnia and Herzegovina, Czech Republic, Estonia, France, Germany, Hungary, Italy, Latvia, Liechtenstein, Lithuania, Poland, Romania, Slovakia, Slovenia, Switzerland, Ukraine	1 year?	>0.5 cm

(continued)

Table 3.2 (continued)

Species	Subfamily	Tribe	Root	Branch	Under bark	Bark-less	Trunk	Alive, declining/dying	Decomposing	Newly fallen/felled	Stored	Timber	Deciduous	Coniferous	Ligneous species infested (non exhaustive list)	European distribution	Cycle	Gallerie diameter
Mesosa curculionides (Linnaeus, 1761)	Lamiinae	Mesosini		x			x			x	x		x		Several deciduous, *Rosaceae*	Balkany region, Austria, Belgium, Czech Republic, France, Germany, Greece, Hungary, Italy, Lithuania, Portugal, Sicily, Spain, Switzerland, The Netherlands, the UK	2–3 years	Around 0.5 cm
Mesosa nebulosa (Fabricius, 1781)	Lamiinae	Mesosini		x	x		x			x	x		x		*Quercus, Castanea, Corylus, Aesculus, Juglans, Tilia, Carpinus, Ilex,* fruit trees …	All Western Europe except Balearic Is., Ireland, Macedonia	2–3 years	Around 0.5 cm
Molorchus minor (Linnaeus, 1758)	Cerambycinae	Molorchini		x	x		x		x	x	x			x	*Abies, Picea, Pinus, Larix*	All Western Europe except Mediterranean is., Albania, Ireland, Macedonia, Spain, Portugal	2 years	>0.5 cm
Monochamus galloprovincialis (Olivier, 1795)	Lamiinae	Monochamini		x	x		x			x				x	*Pinus halepensis, P. maritima, P. sylvestris, P. laricio, Picea Abies*	All Western Europe except Belgium, Ireland, Luxembourg, Sardinia, Sicily, the UK	1 year	Around 1 cm
Monochamus sartor (Fabricius, 1787)	Lamiinae	Monochamini		x	x		x	x		x	x			x	Mainly: *Picea* More rarely: *Abies, Pinus*	Albania, Austria, Bosnia and Herzegovina, Bulgaria, Croatia, Czech Republic, France, Germany, Hungary, Italy, Liechtenstein, Lithuania, Poland, Romania, Slovakia, Slovenia, Switzerland, Ukraine	1–2 years	1–2 cm
Monochamus sutor (Linnaeus, 1758)	Lamiinae	Monochamini		x	x		x	x		x	x			x	Mainly: *Picea* More rarely: *Abies, Pinus, Larix*	All Western Europe except Mediterranean is., Croatia, Greece, Ireland, Luxembourg, Macedonia, Portugal, the UK	1 year	Around 1 cm

Species	Subfamily	Tribe									Host plants	Countries	Development	Size
Morimus asper (Sulzer, 1776)	Lamiinae	Phrissomini	x	x	x	x	x	x	x	x	Mainly: *Fagus* More rarely: *Populus, Alnus, Salix, Quercus, Carpinus, Juglans, Tilia* … *Abies, Cedrus, Larix, Pinus*	Balkany region, Austria, Czech Republic, France, Greece, Hungary, Italy, San Marino, Sardinia, Sicily, Spain, Switzerland, Ukraine	2–3 years	Around 1 cm
Necydalis major (Linnaeus, 1758)	Necydalinae	–	x	x		x	x		x		Diverses	All Western Europe except Mediterranean is., Denmark, Ireland, Macedonia, Portugal, the UK	2–3 years?	>1 cm
Necydalis ulmi (Chevrolat, 1838)	Necydalinae	–			x		x	x		x	Diverses	Balkany region except Albania; Austria, Bulgaria, Czech Republic, France, Germany, Greece, Hungary, Italy, Poland, Romania, Spain, Switzerland, Ukraine	3–4 years	>1 cm
Niphona picticornis (Mulsant, 1839)	Lamiinae	Pteropliini	x		x		x		x	x	Mainly: *Pistacia lentiscus, Genista, Ficus, Cercis siliquastrum* More rarely: *Pistacia terebinthus, Ulmus, Robinia, Laurus, Cistus* …	Balkany region except Macedonia; Mediterranean is., Bulgaria, France, Greece, Italy, Spain	2–3 years	Around 0.5 cm
Nothorhina muricata (Dalman, 1817)	Aseminae	Asemini		x	x			x		x	*Pinus maritima, P. sylvestris, P. Uncinata*	Austria, Bosnia and Herzegovina, Croatia, Czech Republic, Estonia, Finland, France, Germany, Greece, Italy, Latvia, Lithuania, Poland, Portugal, Slovakia, Spanish, Sweden, Ukraine	2 years	>0.5 cm
Nustera distigma (Charpentier, 1825)	Lepturinae	Lepturini	x		x		x	x	x	x	*Pinus*	France, Portugal, Spain	2 years?	>1 cm

(continued)

Table 3.2 (continued)

Species	Subfamily	Tribe	Root	Branch	Under bark	Bark-less	Trunk	Alive, declining/dying	Decomposing	Newly fallen/felled	Stored	Timber	Deciduous	Coniferous	Ligneous species infested (non exhaustive list)	European distribution	Cycle	Gallerie diameter
Obrium brunneum (Fabricius, 1792)	Cerambycinae	Obriini		x	x		x			x	x			x	Mainly: *Picea* More rarely: *Pinus, Larix, Abies*	All Western Europe except Mediterranean is., Albania, Ireland, Macedonia, Portugal	1–2 years	>0.5 cm
Obrium cantharinum (Linnaeus, 1767)	Cerambycinae	Obriini		x	x		x			x	x		x		*Populus, Salix, Rosa, Quercus, Ficus*	All Western Europe except Mediterranean is., Balkany region, Denmark, Ireland, Portugal, Spain	1–2 years	>0.5 cm
Oplosia cinerea (Mulsant, 1839)	Lamiinae	Acanthoderini		x	x			x	x	x			x		Mainly: *Tilia* More rarely: *Fagus, Corylus, Sorbus, Prunus, Fraxinus, …*	Austria, Bosnia and Herzegovina, Croatia, Czech Republic, Denmark, Estonia, Finland, France, Germany, Hungary, Italy, Norway, Poland, Slovakia, Sweden, Switzerland, Ukraine	2 years	>0.5 cm
Oxymirus cursor (Linnaeus, 1758)	Lepturinae	Rhagiini	x	x			x		x	x	x		x	x	*Picea, Pinus, Abies* … Less often: *Fagus, Alnus, Betula*	All Western Europe except Mediterranean is., Albania, Greece, Ireland, Portugal, the UK	2–3 years	>1 cm
Oxypleurus nodieri (Mulsant, 1839)	Aseminae	Saphanini	x	x	x	x	x	x	x	x	x	?		x	*Pinus pinaster, P. nigra, P. salzmannii, P. sylvestris, P. halepensis*	Mediterranean is., Croatia, France, Greece, Italy, Portugal, Spain	2 years minimum (very variable)	>1 cm
Pachyta lamed (Linnaeus, 1758)	Lepturinae	Rhagiini	x		x			x						x	Mainly: *Picea* More rarely: *Pinus*	Austria, Bulgaria, Czech Republic, Denmark, Finland, France, Germany, Hungary, Italy, Latvia, Lithuania, Moldova, Norway, Poland, Romania, Slovakia, Sweden, Switzerland, Ukraine	3 years minimum	>1 cm
Pachyta quadrimaculata (Linnaeus, 1758)	Lepturinae	Rhagiini	x		x				x	x				x	Mainly: *Pinus* More rarely: *Abies, Picea, Larix*	All Western Europe except Mediterranean is., Albania, Belgium, Ireland, Portugal, The Netherlands, the UK	3 years	>1 cm

Species	Subfamily	Tribe										Host plants	Distribution	Generation	Size
Paracorymbia maculicornis (De Geer, 1775)	Lepturinae	Lepturini	x	x		x		x			x	x	Mainly: *Pinus, Picea, Abies* More rarely: *Fagus, Betula, Quercus*	All Western Europe except Mediterranean is., Denmark, Ireland, Macedonia, Portugal, The Netherlands, the UK	2 years	>0.5 cm
Paracorymbia stragulata (Germar, 1824)	Lepturinae	Lepturini	x				x						*Pinus, Abies*	France, Portugal, Spain	2 years	>0.5 cm
Parmena balteus (Linnaeus, 1767)	Lamiinae	Parmenini		x	x	x	x				x	x	Mainly: *Hedera* More rarely: *Tilia, Sambuscus, Morus, Ficus, Malus, Juglans* … *Pinus, Abies, Picea, Thuya* and grasses	France, Germany, Italy, Switzerland	1 year?	>0.5 cm
Parmena meregallii (Sama, 1984)	Lamiinae	Parmenini	x	x	x	x					x	x	*Malus, Pinus, Ilex, Hedera, Rubus*	France, Spain	1 year?	>0.5 cm
Parmena unifasciata (Rossi, 1790)	Lamiinae	Parmenini	x	x	x	x			x		x	x	Diverses	Albania, Bosnia and Herzegovina, Croatia, France, Germany, Greece, Hungary, Italy, Romania, Slovenia, Switzerland	1 year?	>0.5 cm
Pedostrangalia (Etorofus) pubescens (Fabricius, 1787)	Lepturinae	Lepturini	x	x	x	x	x	x			x	x	*Pinus*	Austria, Bosnia and Herzegovina, Bulgaria, Croatia, Czech Republic, France, Germany, Greece, Italy, Poland, Romania, Slovakia, Slovenia, Spain, Switzerland	3 years	>1 cm
Pedostrangalia revestita (Linnaeus, 1767)	Lepturinae	Lepturini	x	x			x	x			x	x	Mainly: *Quercus, Populus, Ulmus* More rarely: *Fagus, Juglans Salix, Acer, Prunus* …	All Western Europe except Balearic Is., Albania, Ireland, Luxembourg, Macedonia	2–3 years	>0.5 cm

(continued)

Table 3.2 (continued)

Species	Subfamily	Tribe	Root	Branch	Under bark	Bark-less	Trunk	Alive, declining/dying	Decomposing	Newly fallen/felled	Stored	Timber	Decid-uous	Conifer-ous	Ligneous species infested (non exhaustive list)	European distribution	Cycle	Gallerie diameter
Penichroa fasciata (Stephens, 1831)	**Cerambycinae**	**Gracilini**		x	x		x			x	x		x	x	**Diverses**	**Balkany region except Albania and Macedonia; Mediterranean is, Belgium, France, Italy, Portugal, Spain**	**2 years**	**>0.5 cm**
Phymatodes testaceus (Linnaeus, 1758)	Cerambycinae	Callidiini			x					x	x		x		Quercus	All Western Europe except Balearic Is., Ireland	1–2 years	>1 cm
Plagionotus arcuatus (Linnaeus, 1758)	Cerambycinae	Clytini			x		x			x	x		x		Quercus, Castanea, Fagus …	All Western Europe except Balearic Is., Ireland, Macedonia, the UK	1 years?	Around 1 cm
Plagionotus detritus (Linnaeus, 1758)	Cerambycinae	Clytini		x	x		x	x		x			x		Mainly: Quercus More rarely: Castanea, Betula, Carpinus, Salix, Fagus	All Western Europe except Balearic Is., Ireland, Luxembourg, Macedonia, Sardinia, Sicily	2 years	Around 1 cm
Poecilium alni (Linnaeus, 1767)	Cerambycinae	Callidiini		x	x			x		x	x		x		Mainly: Quercus More rarely: Castanea, Alnus, Acer, Fraxinus, Ulmus, Corylus, Rosa	All Western Europe except Albania, Balearic Is., Ireland, Macedonia, Crete, Sardinia	1 year?	>0.5 cm
Poecilium fasciatum (Villiers, 1789)	Cerambycinae	Callidiini		x	x		x	x		x	x		x		Vitis vinifera	Balkany region except Albania, Macedonia and Slovenia; Austria, Czech Republic, Mediterranean is, France, Greece, Italy, Hungary, Spain, Switzerland, Ukraine	1 year	>0.5 cm
Poecilium glabratum (Charpentier, 1825)	Cerambycinae	Callidiini		x	x		x	x		x	x			x	Juniperus, Cupressus	Austria, Bosnia and Herzegovina, Croatia, Czech Republic, France, Germany, Greece, Hungary, Italy, Poland, Romania, Sardinia, Slovakia, Spain, Switzerland, Ukraine	1–2 years	>0.5 cm

Species	Subfamily	Tribe								Host plant	Distribution	Development	Ø
Poecilium lividum (Rossi, 1794)	Cerambycinae	Callidiini	x	x		x		x	x	*Quercus ilex, Q. pubescens, Q. robur, Q. suber*	Mediterranean is., Belgium, Bosnia and Herzegovina, Bulgaria, Croatia, Czech Republic, France, Greece, Italy, Portugal, Romania, Slovakia, Spain, Ukraine	1 year?	>0.5 cm
Poecilium pusillum (Fabricius, 1787)	Cerambycinae	Callidiini	x	x	x	x		x	?	Mainly: *Quercus pubescens, Q. robur* More rarely: *Catanea, Fagus*	Austria, Belgium, Bosnia and Herzegovina, Bulgaria, Croatia, Czech Republic, France, Germany, Greece, Hungary, Italy, Moldova, Poland, Romania, Slovakia, Slovenia, Spain, Sweden, Switzerland, Ukraine	1–2 years	>0.5 cm
Poecilium rufipes (Fabricius, 1777)	Cerambycinae	Callidiini	x			x		x	x	Mainly: *Crataegus, Prunus spinoza* More rarely: *Juglans, Quercus, Ulmus, Cornus, Rubus*	Austria, Belgium, Bosnia and Herzegovina, Bulgaria, Croatia, Czech Republic, France, Germany, Greece, Hungary, Italy, Macedonia, Moldova, Poland, Romania, Slovakia, Slovenia, Spain, Switzerland, Ukraine	1 year?	>0.5 cm
Pogonocherus caroli (Mulsant, 1862)	Lamiinae	Pogonocherini	x	x		x		x	x	*Pinus*	France, Spain, Sweden	2 years?	>0.5 cm
Pogonocherus decoratus (Fairmaire, 1855)	Lamiinae	Pogonocherini	x	x		x		x	x	Mainly: *Pinus* Reported: *Abies, Picea, Corylus*	Austria, Belgium, Bosnia and Herzegovina, Bulgaria, Croatia, Czech Republic, Denmark, Estonia, Finland, France, Germany, Greece, Hungary, Italy, Latvia, Lithuania, Norway, Poland, Slovakia, Slovenia, Spain, Sweden, Switzerland, The Netherlands, Ukraine	2 years?	>0.5 cm

(continued)

Table 3.2 (continued)

Species	Subfamily	Tribe	Root	Branch	Under bark	Bark-less	Trunk	Alive, declining/ dying	Decom-posing	Newly fallen/ felled	Stored	Timber	Decid-uous	Conifer-ous	Ligneous species infested (non exhaustive list)	European distribution	Cycle	Gallerie diameter
Pogonocherus fasciculatus (De Geer, 1775)	Laminae	Pogonocherini		x	x			x		x			x	x	Mainly: *Pinus, Picea, Abies, Larix* Reported: *Castanea, Ficus*	All Western Europe except Mediterranean is., Albania, Ireland, Portugal	2 years	>0.5 cm
Pogonocherus hispidulus (Piller and Mitterpacher, 1783)	Laminae	Pogonocherini		x	x			x		x			x		*Tilia* and fruit trees, *Betula, Corylus, Quercus* …	All Western Europe except Balearic Is., Macedonia	1 year	>0.5 cm
Pogonocherus hispidus (Linnaeus, 1758)	Laminae	Pogonocherini		x	x			x		x			x		*Quercus, Tilia, Ulmus, Fagus, Juglans,* Fruit trees …	All Western Europe except Balearic Is., Macedonia, Crete, Sardinia	1 year	>0.5 cm
Pogonocherus ovatus (Goeze, 1777)	Laminae	Pogonocherini		x	x			x		x			x	x	Mainly: *Abies* More rarely: *Pinus, Picea* Reported: *Quercus, Castanea*	Austria, Belgium, Bosnia and Herzegovina, Croatia, Czech Republic, France, Germany, Greece, Hungary, Italy, Latvia, Liechtenstein, Lithuania, Poland, Romania, Slovakia, Slovenia, Spain, Switzerland, The Netherlands, Ukraine	2 years?	>0.5 cm
Pogonocherus perroudi (Mulsant, 1839)	Laminae	Pogonocherini		x	x			x		x				x	*Pinus*	Mediterranean is., Croatia, France, Greece, Italy, Slovenia, Spain, Ukraine	2 years?	>0.5 cm
Prinobius myardi (Mulsant, 1842)	Prioninae	Macrotomini	x	x	x		x	x		x	x		x		Mainly: *Quercus suber; Q. ilex* More rarely: *Fraxinus, Acer, Populus, Alnus, Salix,* fruit trees	Mediterranean is. except Balearic Is., Albania, Bosnia and Herzegovina, Croatia, France, Greece, Italy, Portugal, Spain, Ukraine	1 year?	Until 3 cm
Prionus coriarius (Linnaeus, 1758)	Prioninae	Prionini	x				x		x				x	x	Mainly: *Quercus, Fagus,* More rarely: *Castanea, Fraxinus, Ulmus, Salix, Betula* …	All Western Europe except Balearic Is., Ireland, Crete, Sardinia	2–3 years	2–3 cm

Species	Subfamily	Tribe									Host plants	Distribution	Generation	Size
Pseudovadonia livida (Fabricius, 1777)	Lepturinae	Lepturini	x	x		x		x		x	*Quercus, Castanea, Fagus*	All Western Europe except Balearic Is., Crete, Sardinia	2 years	>0.5 cm
Purpuricenus budensis (Götz, 1783)	Cerambycinae	Purpuricenini	x	x		x	x		x	x	*Pistacia lentiscus, Q. ilex, Salix, Fagus, Ulmus …*	Austria, Balkany region, France, Greece, Italy, Spain	2 years	>1 cm
Purpuricenus globulicollis (Dejean, 1839)	Cerambycinae	Purpuricenini	x	x		x			x	x	Mainly: *Acer* More rarely: *Quercus, Crataegus, Prunus, Rhamnus*	Bosnia and Herzegovina, Bulgaria, Croatia, France, Greece, Italy, Romania, Slovenia, Spain	2 years	>1 cm
Purpuricenus kaehleri (Linnaeus, 1758)	Cerambycinae	Purpuricenini	x	x			x	x	x	x	*Ulmus, Rhamnus, Quercus, Robinia, Prunus …*	All Western Europe except Balearic Is., Denmark, Ireland, Crete, Luxembourg, Sardinia, The Netherlands, the UK	2–3 years	>1 cm
Pyrrhidium sanguineum (Linnaeus, 1758)	Cerambycinae	Callidiini	x	x	x	x		x	x	x	Mainly: *Quercus* More rarely: *Fagus, Betula, Ulmus, Castanea, Carpinus, Malus, Prunus dulcis*	All Western Europe except Balearic Is., Ireland	1–2 years	>1 cm
Rhagium bifasciatum (Fabricius, 1775)	Lepturinae	Rhagiini	x	x			x	x	x	x	*Quercus, Betula, Fagus, Acer, Tilia, Juglans, Castanea, Fraxinus, Alnus …* *Pinus, Abies, Picea, Larix*	All Western Europe except Mediterranean is.	2 years	>1 cm
Rhagium inquisitor (Linnaeus, 1758)	Lepturinae	Rhagiini	x			x	x	x	x	x	Mainly: *Pinus* More rarely: *Abies, Picea, Cedrus, Larix …* Maybe on *Betula, Quercus, Fagus*	All Western Europe except Balearic Is., Ireland, Crete, Sardinia	1 year	Around 1 cm

(continued)

Table 3.2 (continued)

Species	Subfamily	Tribe	Root	Branch	Under bark	Bark-less	Trunk	Alive, declining/ dying	Decomposing	Newly fallen/ felled	Stored	Timber	Deciduous	Coniferous	Ligneous species infested (non exhaustive list)	European distribution	Cycle	Gallerie diameter
Rhagium mordax (De Geer, 1775)	Lepturinae	Rhagiini	x	x	x		x		x	x			x	x	*Quercus, Betula, Fagus, Acer, Tilia, Juglans, Castanea, Fraxinus, Alnus* … *Pinus, Abies, Picea, Larix*	All Western Europe except Mediterranean is., Luxembourg, Portugal	2–3 years	>1 cm
Rhagium sycophanta (Schrank, 1781)	Lepturinae	Rhagiini	x		x		x	x					x		*Quercus, Castanea, Fagus, Alnus, Betula, Prunus*	All Western Europe except Mediterranean is., Luxembourg, the UK	2–3 years	1–1.5 cm
Rhamnusium bicolor (Schrank, 1781)	Lepturinae	Rhagiini		x			x	x					x		Mainly: *Aesculum, Fagus, Acer, Populus, Tilia* More rarely: *Ulmus, Juglans, Salix, Quercus, Robinia*	All Western Europe except Balearic Is., Denmark, Ireland, Crete, Portugal, Sardinia, the UK	1–2 years	>1 cm
Ropalopus clavipes (Fabricius, 1775)	Cerambycinae	Callidiini		x						x	x		x	x	Diverses	Balkany region, Belgium, Denmark, France, Germany, Greece, Hungary, Italy, Crete, Latvia, Lithuania, Poland, Spain, Switzerland, The Netherlands, Ukraine	2 years	1–2 cm
Ropalopus femorantus (Linnaeus, 1758)	Cerambycinae	Callidiini		x	x					x	x		x	x	Diverses	Austria, Belgium, Bosnia and Herzegovina, Bulgaria, Croatia, Czech Republic, France, Germany, Greece, Hungary, Italy, Latvia, Macedonia, Romania, Slovakia, Slovenia, Spain, Sweden, Switzerland, Ukraine	2 years	Around 1 cm

Species	Subfamily	Tribe									Host plants	Geographic distribution	Generation	Size
Ropalopus insubricus (Germar, 1824)	Cerambycinae	Callidiini		x	x	x				x	x	Mainly: *Acer* More rarely: *Fraxinus, Alnus, Fagus, Salix*	Austria, Bosnia and Herzegovina, Bulgaria, Croatia, Czech Republic, France, Greece, Italy, Macedonia, Romania, Slovakia, Slovenia, Spain, Ukraine	2 years	1–2 cm
Ropalopus ungaricus (Herbst, 1784)	Cerambycinae	Callidiini		x	x	x				x	x	*Acer pseudoplatanus*	Austria, Bosnia and Herzegovina, Bulgaria, Croatia, Czech Republic, France, Germany, Hungary, Italy, Poland, Romania, Slovakia, Slovenia, Switzerland, Ukraine	2 years	1–2 cm
Ropalopus varini (Bedel, 1870)	Cerambycinae	Callidiini		x	x	x				x	x	*Quercus, Q. ilex*	Austria, Belgium, Croatia, Czech Republic, France, Germany, Greece, Hungary, Italy, Macedonia, Moldova, Romania, Slovakia, Slovenia, Spain, Switzerland, Ukraine	1 year minimum	Around 1 cm
Rosalia alpina (Linnaeus, 1758)	Cerambycinae	Rosaliini		x	x	x	x	x		x	x	Mainly: *Fagus* More rarely: *Fraxinus, Alnus, Tilia, Acer, Carpinus*	Balkany region, Austria, Czech Republic, France, Germany, Greece, Hungary, Italy, Liechtenstein, Poland, Sicily, Spain, Switzerland, Ukraine	2–3 years minimum	>2 cm
Rutpela maculata (Poda, 1761)	Lepturinae	Lepturini	x	x	x			x		x	x	*Quercus, Alnus, Crataegus, Pyrus, Populus, Fagus, Salix, Genista …*	All Western Europe except Balearic Is., Crete	2 years	>1 cm
Saperda carcharias (Linnaeus, 1758)	Lamiinae	Saperdini		x	x	x				x	x	*Populus*	All Western Europe except Mediterranean is., Ireland, Portugal	3 years	1–2 cm

(continued)

Table 3.2 (continued)

Species	Subfamily	Tribe	Root	Branch	Under bark	Bark-less	Trunk	Alive, declining/dying	Decom-posing	Newly fallen/felled	Stored	Timber	Decid-uous	Conifer-ous	Ligneous species infested (non exhaustive list)	European distribution	Cycle	Gallerie diameter
Saperda octopunctata (Scopoli, 1772)	Lamiinae	Saperdini		x	x		x	x					x		*Tilia* More rarely: *Populus*	Balkany region except Macedonia; Austria, Belgium, Czech Republic, France, Germany, Greece, Hungary, Italy, Poland, Spain, Switzerland, The Netherlands, Ukraine	1–2 years	>1 cm
Saperda perforata (Pallas, 1773)	Lamiinae	Saperdini		x	x		x	x					x		*Salix, Populus*	Austria, Bosnia and Herzegovina, Bulgaria, Croatia, Czech Republic, Estonia, Finland, France, Germany, Hungary, Italy, Latvia, Lithuania, Moldova, Norway, Poland, Romania, Slovakia, Sweden, Switzerland, The Netherlands, Ukraine	1–2 years?	>1 cm
Saperda populnea (Linnaeus, 1758)	Lamiinae	Saperdini		x	x			x					x		*Populus*	All Western Europe except Balearic Is., Ireland, Crete	1–2 years	>1 cm
Saperda punctata (Linnaeus, 1767)	Lamiinae	Saperdini		x	x		x	x					x		Mainly: *Ulmus* More rarely: *Quercus, Tilia*	Balkany region, Mediterranean is. except Crete; Austria, Czech Republic, France, Germany, Greece, Hungary, Italy, Latvia, Poland, Spain, Ukraine	1–2 years	>1 cm
Saperda scalaris (Linnaeus, 1758)	Lamiinae	Saperdini	x	x	x		x	x					x		*Cerasus*, other fruit trees	All Western Europe except Balearic Is., Ireland, Crete, Portugal, Sardinia	1–3 years	>1 cm

Species	Subfamily	Tribe									Host plants	Distribution	Development time	Size
Saperda similis (Laicharting, 1784)	Lamiinae	Saperdini	x	x	x	x	x			x	*Salix caprea, S. purpurea*	Albania, Austria, Belgium, Bulgaria, Croatia, Czech Republic, Estonia, Finland, France, Germany, Greece, Hungary, Italy, Latvia, Norway, Poland, Romania, Slovakia, Slovenia, Spain, Sweden, Switzerland, Ukraine	2–3 years	Around 1 cm
Saphanus piceus (Laicharting, 1784)	Aseminae	Saphanini	x	x	x	x	x	x	x	x	Mainly: *Corylus* More rarely: *Salix, Fagus, Betula, Carpinus, Aubépine, Quercus, Castanea…* Very rarely: *Abies, Picea*	Austria, Bosnia and Herzegovina, Bulgaria, Croatia, Czech Republic, France, Germany, Greece, Hungary, Italy, Macedonia, Poland, Romania, Slovakia, Sweden, Switzerland	3 years minimum	Around 1 cm
Semanotus laurasii (Lucas, 1852)	Cerambycinae	Callidiini	x	x	x					x	*Juniperus*	France, Sardinia, Spain	2 years	Around 1 cm
Semanotus undatus (Linnaeus, 1758)	Cerambycinae	Callidiini	x	x	x	x				x	Mainly: *Picea, Abies* More rarely: *Pinus*	Austria, Bosnia and Herzegovina, Bulgaria, Croatia, Czech Republic, Estonia, Finland, France, Germany, Hungary, Italy, Latvia, Liechtenstein, Lithuania, Norway, Poland, Romania, Slovakia, Slovenia, Sweden, Switzerland, Ukraine	1 year minimum	Around 1 cm
Spondylis buprestoides (Linnaeus, 1758)	Aseminae	Spondylidinae	x	x	x	x	x	x	x	x	*Pinus, Abies, Picea*	All Western Europe except Albania, Balearic Is., Ireland, Crete, Sardinia, Switzerland, the UK	1 year?	Around 1 cm
Stenocorus meridianus (Linnaeus, 1758)	Lepturinae	Rhagiini	x	x	x	x				x	*Ulmus, Quercus, Fagus, Fraxinus, Acer, Malus, Prunus*	All Western Europe except Mediterranean is., Ireland, Portugal	2–3 years	>1 cm

(continued)

Table 3.2 (continued)

Species	Subfamily	Tribe	Root	Branch	Under bark	Bark-less	Trunk	Alive, declining/dying	Decom-posing	Newly fallen/felled	Stored	Timber	Decid-uous	Conifer-ous	Ligneous species infested (non exhaustive list)	European distribution	Cycle	Gallerie diameter
Stenopterus ater (Linnaeus, 1767)	Cerambycinae	Stenopterini		x			x			x	x		x		*Pistacia, Cercis siliquastrum, Corylus, Eucalyptus* …	Balearic Is., Croatia, France, Italy, Macedonia, Portugal, Sardinia, Sicily, Slovakia, Slovenia, Spain, Switzerland, Ukraine	2 years	>0.5 cm
Stenopterus rufus (Linnaeus, 1767)	Cerambycinae	Stenopterini		x			x			x	x		x		*Quercus ilex, Q. coccifera, Castanea, Robinia, Juglans, Prunus, Acacia, Pistacia, Ostrya, Salix, Rhamnus*	All Western Europe except Ireland, Crete, Luxembourg, the UK	2 years	>0.5 cm
Stenostola dubia (Laicharting, 1784)	Laminae	Saperdini		x	x			x		x	x		x		*Tilia, Quercus, Salix, Juglans, Corylus, Alnus, Betula, Populus, Carpinus,* fruit trees …	Austria, Belgium, Bosnia and Herzegovina, Croatia, Czech Republic, Denmark, France, Germany, Hungary, Italy, Latvia, Liechtenstein, Luxembourg, Moldova, Norway, Poland, Romania, Slovakia, Slovenia, Spain, Sweden, Switzerland, The Netherlands, Ukraine	2 years	>0.5 cm
Stenostola ferrea (Schrank, 1776)	Laminae	Saperdini		x	x			x		x	x		x		Mainly: *Tilia* More rarely: *Salix, Juglans, Corylus, Ulmus, Fagus, Quercus, Castanea*	Austria, Belgium, Bosnia and Herzegovina, Bulgaria, Croatia, Czech Republic, Denmark, Estonia, Finland, France, Germany, Hungary, Italy, Lithuania, Luxembourg, Moldova, Norway, Poland, Romania, Slovakia, Slovenia, Sweden, Switzerland, The Netherlands, the UK, Ukraine	2 years	>0.5 cm

Species	Subfamily	Tribe									Host plants	Distribution	Generation	Size
Stenurella bifasciata (Müller, 1776)	Lepturinae	Lepturini	x	x		x			x	x	Diverses	All Western Europe except Balearic Is., Ireland, Crete, the UK	2 years	>0.5 cm
Stenurella melanura (Linnaeus, 1758)	Lepturinae	Lepturini	x	x		x			x	x	*Quercus, Ulmus, Ficus, Rosa, Pinus, Abies*	All Western Europe except Mediterranean is., Ireland	2 years	>0.5 cm
Stenurella nigra (Linnaeus, 1758)	Lepturinae	Lepturini	x	x		x			x	x	*Quercus, Fagus, Carpinus, Ulmus, Betula, Rhamnus, Corylus, Rosa*	All Western Europe except Mediterranean is., Ireland	2 years minimum	>0.5 cm
Stenurella sennii (Sama, 2002)	Lepturinae	Lepturini	x	x		x			x	x	*Quercus, Ulmus, Ficus, Eglantier, Rosa, Pinus, Abies*	France	2 years	>0.5 cm
Stictoleptura cordigera (Fuesslins, 1775)	Lepturinae	Lepturini	x	x		x	x	x	x	x	*Quercus, Pistacia, Alnus, Q. suber …, Pinus halepensis*	Austria, Belgium, Bosnia and Herzegovina, Bulgaria, Croatia, Czech Republic, France, Germany, Greece, Italy, Luxembourg, Macedonia, Mediterranean is., Romania, Slovakia, Slovenia, Spain, Switzerland, Ukraine	3 years?	Around 1 cm
Stictoleptura erythroptera (Hagenbach, 1822)	Lepturinae	Lepturini		x		x		x	x	x	*Q. rouvre, Q. suber, Fagus, Tilia, Acer, Ulmus*	Austria, Bosnia and Herzegovina, Bulgaria, Croatia, Czech Republic, France, Germany, Greece, Hungary, Romania, Slovakia, Slovenia, Spain, Switzerland, Ukraine	3 years minimum	Around 1 cm
Stictoleptura fontenayi (Mulsant, 1839)	Lepturinae	Lepturini	x	x		x	x	x	x	x	*Quercus, Salix, Ulmus …, Cedrus*	France, Portugal, Spain	2 years	Around 1 cm
Stictoleptura rubra (Linnaeus, 1758)	Lepturinae	Lepturini	x	x		x	x	x	x	x	*Pinus, Abies, Picea, Larix …, Quercus, Fagus, Betula*	All Western Europe except Albania, Ireland, Macedonia, Mediterranean is.	2 years minimum	Around 1 cm

(continued)

Table 3.2 (continued)

Species	Subfamily	Tribe	Root	Branch	Under bark	Bark-less	Trunk	Alive, declining/dying	Decomposing	Newly fallen/felled	Stored	Timber	Deciduous	Coniferous	Ligneous species infested (non exhaustive list)	European distribution	Cycle	Gallerie diameter
Stictoleptura scutellata (Fabricius, 1781)	Lepturinae	Lepturini		x			x	x	x	x			x		Mainly: *Fagus* et *Quercus* More rarely: *Castanea, Carpinus, Corylus, Betula, Alnus*	All Western Europe except Balearic Is., Crete, The Netherlands	2 years?	Around 1 cm
Stictoleptura trisignata (Fairmaire, 1852)	Lepturinae	Lepturini		x	x		x	x					x		*Quercus, Q. suber, Ulmus, Castanea*	Balearic Is., France, Portugal, Spain	2 years?	Around 1 cm
Strangalia attenuata (Linnaeus, 1758)	Lepturinae	Lepturini		x			x		x	x	x		x		*Quercus, Salix, Alnus, Populus, Betula, Corylus, Crataegus, Euonymus, Genista …*	All Western Europe except Ireland, Mediterranean is., Portugal, the UK	2 years	>1 cm
Stromatium unicolor (Olivier, 1795)	Cerambycinae	Hesperophanini		x			x				x	x	x	x	Deciduous: Diverses Coniferous: *Cedrus*	Balkany region, Mediterranean is., France, Greece, Italy, Portugal, Spain	2 years?	1–2 cm
Tetropium castaneum (Linnaeus, 1758)	Aseminae	Asemini	x		x			x		x				x	*Abies, Picea, Pinus*	All Western Europe except Albania, Greece, Macedonia, Ireland, Mediterranean is., Portugal	2 years	Around 1 cm
Tetropium fuscum (Fabricius, 1787)	Aseminae	Asemini	x	x	x		x	x		x	x			x	*Larix* More rarely: *Pinus*	Austria, Belgium, Bosnia and Herzegovina, Bulgaria, Czech Republic, Denmark, Estonia, Finland, France, Germany, Hungary, Italy, Latvia, Liechtenstein, Lithuania, Moldova, Norway, Poland, Romania, Slovakia, Slovenia, Sweden, Switzerland, The Netherlands, Ukraine	1–2 years	Around 1 cm

Species	Subfamily	Tribe									Host plant	Countries	Development	Size
Tetropium gabrieli (Weise, 1905)	Aseminae	Asemini	x	x	x	x	x	x	x	x	*Larix* Very rarely: *Pinus*	Austria, Belgium, Czech Republic, Denmark, France, Germany, Hungary, Italy, Liechtenstein, Luxembourg, Poland, Slovakia, Slovenia, Switzerland, The Netherlands, the UK, Ukraine	1 year	Around 1 cm
Tetrops praeustus (Linnaeus, 1758)	Lamiinae	Tetropini	x	x	x		x	x	x		*Rosaceae* …	All Western Europe except Balearic Is., Ireland, Crete	1 year?	>0.5 cm
Tetrops starkii (Chevrolat, 1859)	Lamiinae	Tetropini	x	x	x		x	x	x		*Fraxinus*	Austria, Bosnia and Herzegovina, Bulgaria, Croatia, Czech Republic, France, Germany, Hungary, Italy, Lithuania, Norway, Poland, Romania, Slovakia, Slovenia, Sweden, the UK, Ukraine	1 year?	>0.5 cm
Tragosoma depsarium (Linnaeus, 1767)	Prioninae	Meroscelisini	x			x	x			x	*Pinus uncinata, Larix, Picea, Abies*	Austria, Bosnia and Herzegovina, Bulgaria, Croatia, Czech Republic, Estonia, Finland, France, Germany, Greece, Italy, Latvia, Liechtenstein, Lithuania, Norway, Poland, Slovakia, Slovenia, Spain, Sweden, Switzerland	3 years, sometimes more	Around 2 cm
Trichoferus fasciculatus (Faldermann, 1837)	Cerambycinae	Hesperophanini	x	x	x		x	x	x		Diverses	Bosnia and Herzegovina, Bulgaria, Croatia, France, Greece, Italy, Mediterranean is., Portugal, Romania, Spain, Ukraine	2 years minimum	Around 1 cm
Trichoferus griseus (Fabricius, 1792)	Cerambycinae	Hesperophanini	x	x	x		x	x	x		*Ficus*	Bulgaria, Croatia, France, Greece, Italy, Mediterranean is., Portugal, Spain, Ukraine	2 years minimum	Around 1 cm

(continued)

Table 3.2 (continued)

Species	Subfamily	Tribe	Root	Branch	Under bark	Bark-less	Trunk	Alive, declining/dying	Decomposing	Newly fallen/felled	Stored	Timber	Deciduous	Coniferous	Ligneous species infested (non exhaustive list)	European distribution	Cycle	Gallerie diameter
Trichoferus holosericeus (Rossi, 1790)	Cerambycinae	Hesperophanini		x	x		x				x	?	x		Quercus ilex, Ficus, Juglans, Fagus, Salix, Ostrya, Robinia, Acacia dealbatta, fruitiers	Mediterranean rim and islands, except Crete	2 years?	1–1.5 cm
Trichoferus pallidus (Olivier, 1790)	Cerambycinae	Hesperophanini		x	x		x	x		x	x	?	x		Mainly: Quercus More rarely: Tilia, Fagus	Austria, Bosnia and Herzegovina, Bulgaria, Croatia, Czech Republic, France, Germany, Greece, Hungary, Poland, Romania, Slovakia, Spain, Ukraine	2 years?	1–1.5 cm
Xylotrechus antilope (Schönherr, 1817)	Cerambycinae	Clytini					x	x		x	x		x		Mainly: Quercus More rarely: Castanea, Fagus, Malus, Prunus mahaleb	All Western Europe except Balearic Is., Denmark, Ireland, Crete, Luxembourg, Macedonia, Sardinia, The Netherlands	1 year?	>1 cm
Xylotrechus arvicola (Olivier, 1795)	Cerambycinae	Clytini		x	x	x	x	x		x	x	x	x		Mainly: Quercus More rarely: Ostrya, Carpinus, Ulmus, Castanea, Fagus, Populus, Salix, Tilia, Crataegus, Prunus …	All Western Europe except Denmark, Ireland, Crete, the UK	Max 2 years	Around 1 cm
Xylotrechus chinensis (Chevrolat, 1852)	Cerambycinae	Clytini		x	x		x	x		x	x		x		Malus, Pyrus, Vitis vinifera	EXOGENOUS: France, Greece, Spain	2 years?	Around 1 cm

Species	Subfamily	Tribe								Host plant	Distribution	Development	Size
Xylotrechus pantherinus (Savenius, 1825)	Ceramby-cinae	Clytini	x	x	x				x	*Salix caprea*	Austria, Czech Republic, Finland, France, Germany, Hungary, Italy, Norway, Poland, Romania, Slovakia, Sweden, Switzerland	2 years	Around 1 cm
Xylotrechus rusticus (Linnaeus, 1758)	Ceramby-cinae	Clytini	x	x	x	x	x	x	x	*Populus, Betula, Salix*	All Western Europe except Balearic Is., Ireland, Crete, Luxembourg, Sicily, The Netherlands, the UK	2 years	Around 1 cm
Xylotrechus stebbingi (Gahan, 1906)	**Ceramby-cinae**	**Clytini**	x	x	x	x	x	x	x	**Softwood deciduous trees: Alnus, Populus ... But also Quercus dilatata**	**EXOGENOUS: France, Greece, Italy, Crete, Sardinia, Switzerland**	**1–2 years**	**Around 1 cm**

Callidium coriaceum (Paykull, 1800)

Capricorn Beetle

Description

The insect is uniformly bronze-brown with greenish or purplish reflections and is 8–16 mm long. Only the antennae, tarsus and tibia apex are brown.

The thorax (pronotum) has a shiny, longitudinal, flat part that is widened at the back and connected to two shorter, equally shiny lateral reliefs.

The elytra are formed at the base of a strong, spaced punctuation which continues backwards with a finer, denser punctuation. Males have sturdy legs with claviform femurs, more so than females.

Distribution

Mainly central Europe: Austria, Belarus, Bosnia and Herzegovina, the Czech Republic, Estonia, Finland, France, Germany, Greece, Hungary, Italy, Latvia, Liechtenstein, Lithuania, Luxembourg, Norway, Poland, Romania, Slovakia, Slovenia, Sweden, Switzerland, Ukraine.

Bioecology

It is a rare species because it is mainly found in the mountains. Adults can be found from June to August looking for a place to lay their eggs on the bark of coniferous trees.

Cycle of Development

Full development usually takes 2 years, but the cycle can be completed in 1 year.

Infested Woods

It is considered as an obligate parasite of spruce (*Picea*) but has been reported on fir (*Abies*), larch (*Larix*), and pine (*Pinus*). It infests wood under the bark and bores into the sapwood of standing weakened/recently fallen/freshly felled trees.

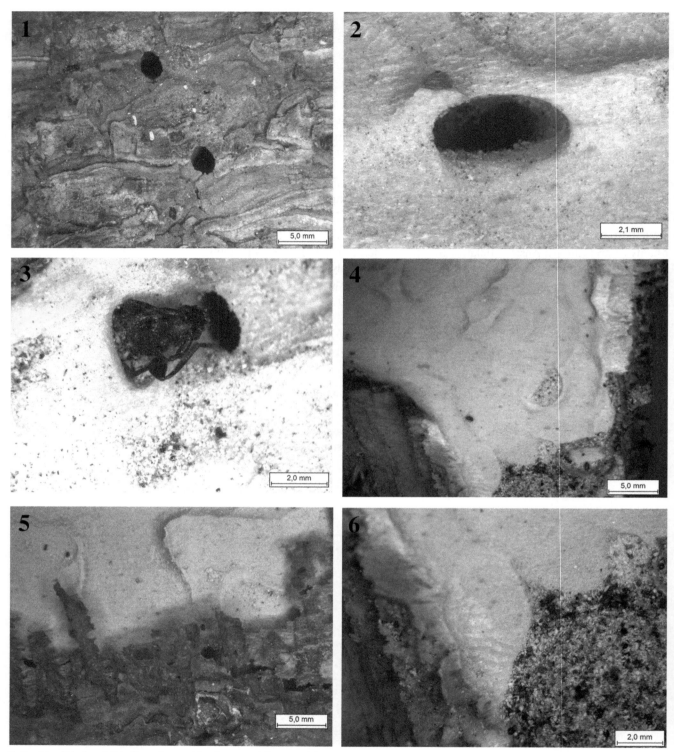

1. Ovoid exit holes visible from the outside of cypress (*Cupressus*) bark
2. Oval larval gallery of *C. coriaceum* located in the sapwood
3. Exit hole cleared from its bark in the laboratory, with imago
4–5. Cambial galleries bored by *C. coriaceum*
6. Mandible traces in the sapwood, detail

The larval galleries of *C. coriaceum* are oval, elliptical, and tend to round off when the imago is formed (Exit holes). The galleries are about 1 cm in diameter, usually 0.5 cm. This species, like many fresh-wood capricorn beetles of conifers, burrows under the bark near resin pockets and frequently penetrates into the sapwood.

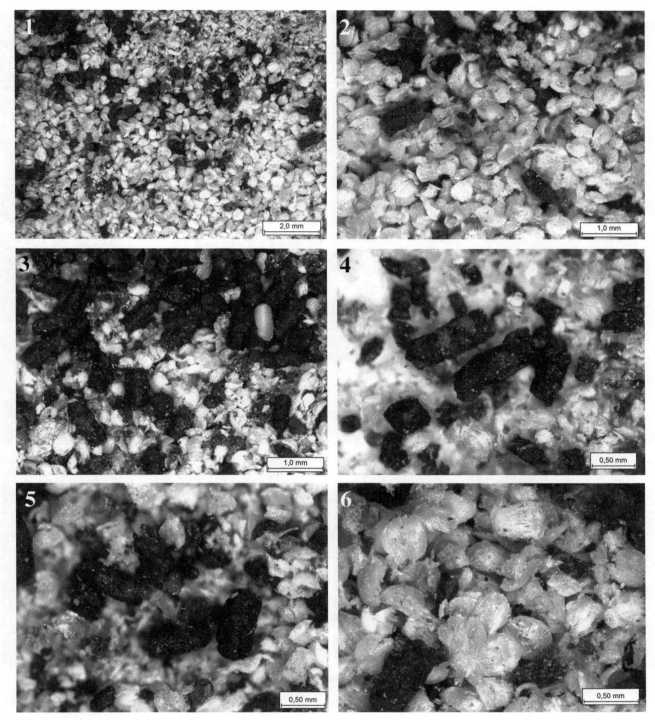

1. *C. coriaceum* frass, general view
2–3. Heterogeneous frass of *C. coriaceum* with a lot of wood chips and small "barrel" pellets
4–5. "Barrel" pellets, detail
6. Wood chips, detail

The frass is heterogeneous: about 20–30% of it is composed of tiny, frail and crumbly cylindrical faecal pellets called "barrels" (<1 mm long and 0.4 mm wide on average) which take on the colour of the cambium. These pellets are mixed with a multitude of small convex chips (0.3–0.4 mm wide) whose colour is closer to that of the sapwood than to that of the cambium. These wood chips are produced by the larva's mandibles; they make up most of the visible sawdust expelled from the gallery by the larva. The "barrels" have a smoother surface than those of *Icosium tomentosum*.

The frass of *C. coriaceum* can be mistaken for the frass of other fresh-wood cerambycids whose frass is not inventoried yet. Identification requires to cross-reference all the characteristics of the woody species with those of the frass in SEM.

1. Frass of *Callidium coriaceum* on *Picea*, several visible friable cylindrical pellets, ×200
2. Two "barrels" with a smooth surface, detail, ×160
3. One "barrel" with smooth agglomerates of various visible digested wood fibres, detail, ×350
4. Zoom on a "barrel" and its smooth surface, ×350
5. Convex, undigested wood chip, with clearly visible wood fibres, detail, ×350
6. Zoom on some undigested tracheids from a cylindrical pellet, ×3500

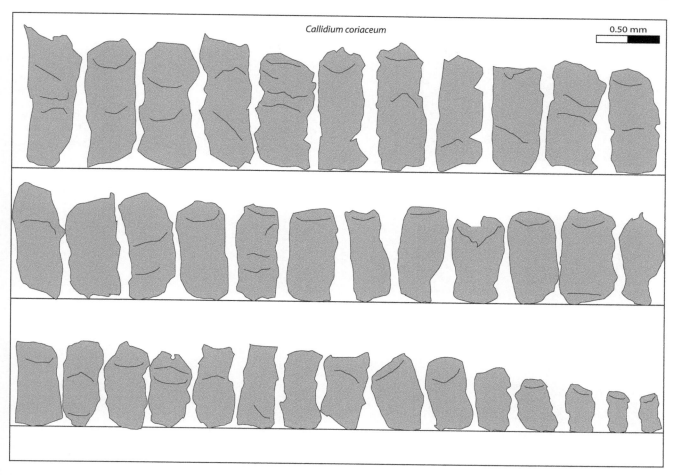

Drawings representing the cylindrical faecal pellets composing the frass of *Callidium coriaceum*. Arranged from the largest to the smallest pellet, collected from a spruce (*Picea*) sample.

Cerambyx cerdo (Linnaeus, 1758)

Great Capricorn Beetle

Description

The size of this species is impressive. Its length varies from 24 to 62 mm, with an extreme length of 120 mm for some individuals. *Cerambyx cerdo* is currently one of the largest wood-boring longhorn beetles in France and Europe.

It is black; its elytra are blackish-brown at their base and gradually turn reddish-brown at their apex.

Its thorax (pronotum) has deep, irregular, more or less rough, strongly entangled wrinkles.

Its elytra are shiny, with granular wrinkles at their base that grow finer in their apical part.

Distribution

This species occurs throughout Western Europe except Denmark, Ireland and Crete. It is a rare species in some places.

In France, it is found in the oak forests of the south and more generally in large forests such as those of Fontainebleau, Compiègne … It seems to be rarer in the north and the center of the country.

The insect is currently protected by law as an endangered species. But although it does tend to become rare in certain regions, it can be found in large numbers in some places and become more threatening than threatened.

Bioecology

The insect comes out at twilight and is first seen in early June, sometimes as early as the end of May. It flies slowly, with its body tilted, its elytra raised and its antennae widely spread in a semi-circle.

The adult's lifetime is short and lasts 1–2 months. It can feed on wounds or various exudations from trees, and even on ripe fruit.

Visible along the trunks or main branches, *C. cerdo* emits a kind of rhythmic squeaking or creaking, produced by repeated rubbing of the thorax on the abdominal area next to it. The eggs are laid singly in cracks and wounds in trees.

The larvae hatch a few days later. The first year the larvae remain in the cortical area. In the second year, they burrow into the wood (sapwood) where they bore sinuous galleries.

At the end of the last instar (summer-autumn), the larva builds an outwardly open gallery and then a pupal chamber that it seals with a calcareous cap.

Cycle of Development

The development cycle lasts at least 2–3 years, sometimes much longer, and depends more particularly on climatic conditions and latitude.

Infested Woods

It mainly attacks the large branches and trunks of oak, other tree species excluded. It needs large enough pieces of wood to grow. It can be found on old, large-diameter trees that are still alive, freshly felled or have been in storage for one season or more. Larvae develop on oak: *Quercus robur*, *Q. petraea*, *Q. pubescens*, *Q. ilex* and *Q. suber*. It has also been reported on chestnut (*Castanea*), ash (*Fraxinus*), elm (*Ulmus*) and alder (*Alnus*).

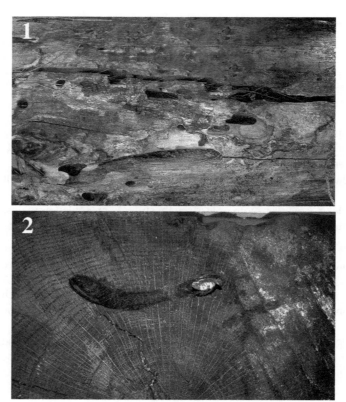

1. Elongated *C. cerdo* galleries on oak trunks felled in the forest (Var "département", France)
2. Example of a gallery attributable to *Cerambyx cerdo* on archaeological waterlogged wood. Maximum diameter 4.5 cm. Oak (*Quercus*) roof frame web, archaeological site of Rezé Saint-Lupien, 2016 (Loire-Atlantique, France)

The galleries of *Cerambyx cerdo* are oval or elliptical to rectangular in shape. They are easily recognizable from their large size. They measure 3 cm in diameter or more on average depending on the individuals. The large size of the galleries and their shape, together with the infested species, provide enough information to avoid confusion with other European woodborers, all of which are much smaller.

Hylotrupes bajulus (Linnaeus, 1758)

House Longhorn Beetle

Description
Black to brown and varying from 7 to 21 mm and up to 25 mm in length, *H. bajulus* is an elongated insect.

Its filiform antennae are composed of 11 items and are much shorter than its body, unlike those of many other longhorned beetles.

Its heart-shaped prothorax has two flat, shiny dorsal protuberances. Its elytra are rough and have vaguely V-shaped bands and patches of silvery whitish bristles.

Distribution
This insect seems to have been rather rare until the nineteenth century, when it found refuge in the new techniques of industrial carpentry construction. Together with termites, *H. bajulus* is currently the number one common pest of coniferous timber in Europe.

H. bajulus is common in almost all of Europe; it is also found on the North American mainland, in North and South Africa and in Asia.

In France, although it is present everywhere, it seems to prefer the temperate climate of the Mediterranean and Atlantic coasts.

Bioecology
Adult emergence usually occurs from June to August. Adults live 1–2 weeks and do not feed, unlike *C. cerdo* adults. They also have poor flying power, they often crawl on the ground.

After a short mating period (3 min), oviposition consists of about 30 eggs on average (sometimes up to 200) and takes place in cracks or crevices in the wood or sometimes in old woodworm galleries.

Very dry wood stored in well-heated places is particularly attractive to this insect, which prefers its soft parts (sapwood). Then the larvae preferably evolve in wood with a humidity rate of less than 30% and with an ambient temperature of 15–28 °C.

The mandibles of the house longhorn beetle are very powerful: while boring galleries, the larvae can pierce a 1.5 mm thick metal plate (which can be found on carpentry works). They can also injure each other if they meet.

Pupation lasts about 2 weeks and takes place in a cavity close to the surface of the wood. Once an adult, the house longhorn beetle reaches the outside world by boring an oval passage (exit hole) through the wood.

Cycle of Development
Two years or more depending on the ambient conditions of the wood being worked.

The development period generally lasts 3–6 years and can reach 10 years in unfavourable conditions (ambient temperature below 15 °C).

Infested Woods
The house longhorn beetle grows best in coniferous trees. It is particularly fond of decaying or dead wood as well as worked wood (frames, parquet flooring, beams, joists, etc.). It is the most destructive capricorn beetle for timber and heritage woods. It does not need bark to develop.

Nevertheless, it can adapt to a few deciduous trees such as tamarisk (*Tamarix*) and poplar (*Populus*).

Pines (*Pinus*) (maritime, Scots, Austrian black …) are particularly infested; fir (*Abies*), spruce (*Picea*), larch (*Larix*), and Douglas fir (*Pseudotsuga menziesii*) are less so. The strongest recorded infestations occur during the first 20–40 years after felling.

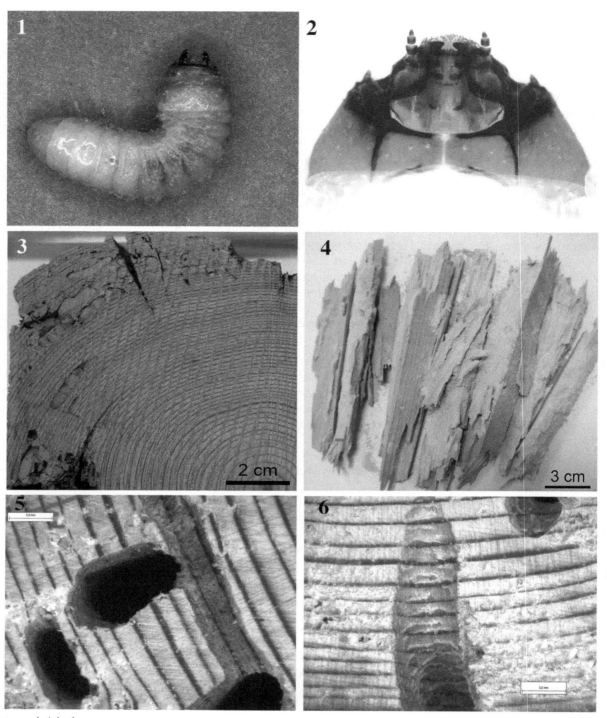

1. *Hylotrupes bajulus* larva
2. Larva head and mandibles, detail
3. Sample of infested larch (*Larix*), general view
4. Sample of infested pine (*Pinus sylvestris*), general view. Important colonisation leaving only a thin layer of the wood (final wood most often), as thin as paper
5. Oval galleries of *H. bajulus* in pine tree (*Pinus sylvestris*)
6. Traces left by *H. bajulus* mandibles, detail

The galleries of *H. bajulus* are about 1 cm in diameter, oval to elliptical, and usually found in the sapwood of conifers. Infestation can only be detected after the first generation has come out. If the wood is already heavily infested, this species can penetrate into the heartwood.

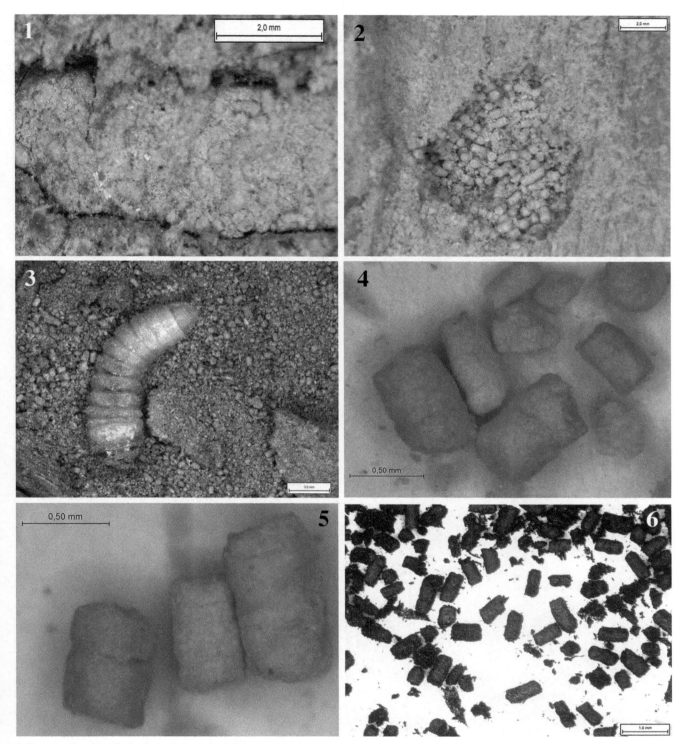

1. Faecal pellets in galleries in larch (*Larix*); visible compacting
2. Faecal pellets in galleries in Scots pine (*Pinus sylvestris*)
3. *H. bajulus* larva and frass, Scots pine (*Pinus sylvestris*)
4–5. "Barrels" in larch (*Larix*), detail
6. Experimentally carbonised "barrels", Scots pine (*Pinus sylvestris*), detail

The frass of *H. bajulus* is homogenous and consists of small cylindrical faecal pellet, called "barrels" of variable sizes, around 1 mm in length. These pellets have a strong tendency to crumble, and make the whole frass look like powder. This aspect is the result of a partial digestion of the wood fibres which sometimes agglomerate in an anarchic way.

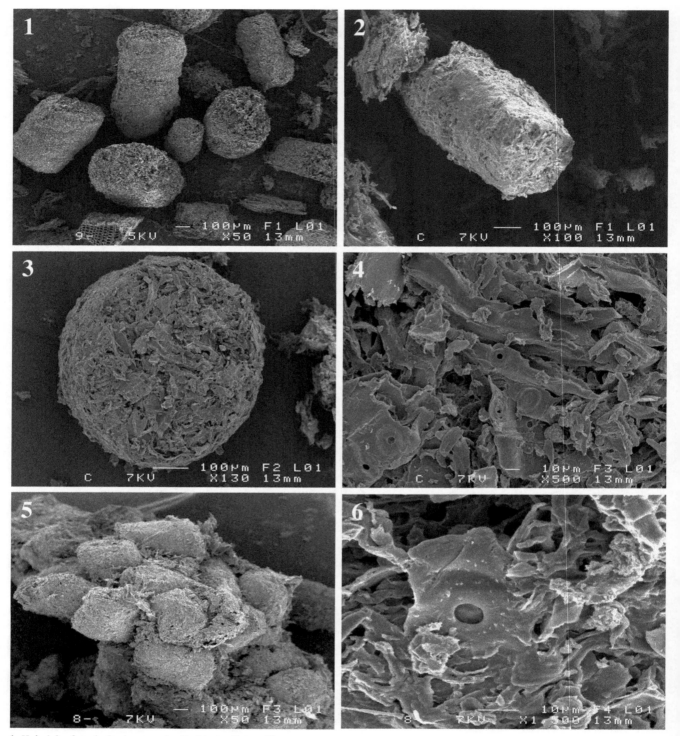

1. *H. bajulus* frass in *Larix*, several cylindrical pellets, ×50
2. "Barrel" in *Larix,* detail, ×100
3. Top view of a "barrel", detail. Agglomerate of various wood fibres visible, ×130
4. Zoom on undigested *Pinus* areola-like punctuations, visible within a pellet, ×500
5. Experimentally carbonised faecal pellet of *H. bajulus* in *Pinus sylvestris*, CReAAH, ×50
6. Zoom on an experimentally carbonised, undigested, areola-like punctuation in *Pinus*, CReAAH, ×1500

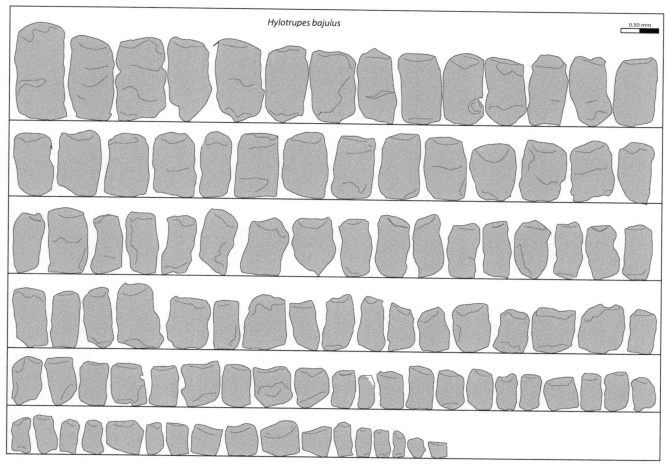

Drawings representing 100 or so cylindrical faecal pellet composing the frass of *Hylotrupes bajulus*. Arranged from the largest to the smallest pellet, harvested from the larch (*Larix*) sample mentioned above.

Icosium tomentosum (Lucas, 1854)

Capricorn Beetle

Description

Reddish brown to dark brown and 8–16 mm long, this species has elongated, parallel elytra with prominent shoulders rounded at the apex and long, fairly sturdy legs.

The thorax (pronotum) is longer than wide and narrower at its front end than at its rear end, and bears three parallel, smooth, narrow bands.

This species is divided into two subspecies: *I. tomentosum tomentosum* and *I. tomentosum atticum* (Sama, 1988).

Distribution

Mediterranean species: Balearic Islands, Bosnia and Herzegovina, Croatia, Cyprus, France, Greece, Italy, Crete, Malta, Sardinia, Spain.

Bioecology

Adults are found from June to September. They rather come out at twilight and at night. The female lays its eggs in cracks in the wood under the bark of dead trunks or small branches of dying or freshly fallen or felled trees. Optimal temperatures for larval development are between 18 and 22 °C.

Cycle of Development

2 years.

Infested Woods

Exclusively on coniferous trees and more particularly in the cambium of trees of the Cupressaceae family: juniper (*Juniperus*), (thuja) *Thuja*, cypress (*Cupressus* and *Tetraclinis* spp.). The larva remains under the bark during the first part of its life and then bores into the sapwood where it establishes its pupation chamber. Can penetrate the sapwood.

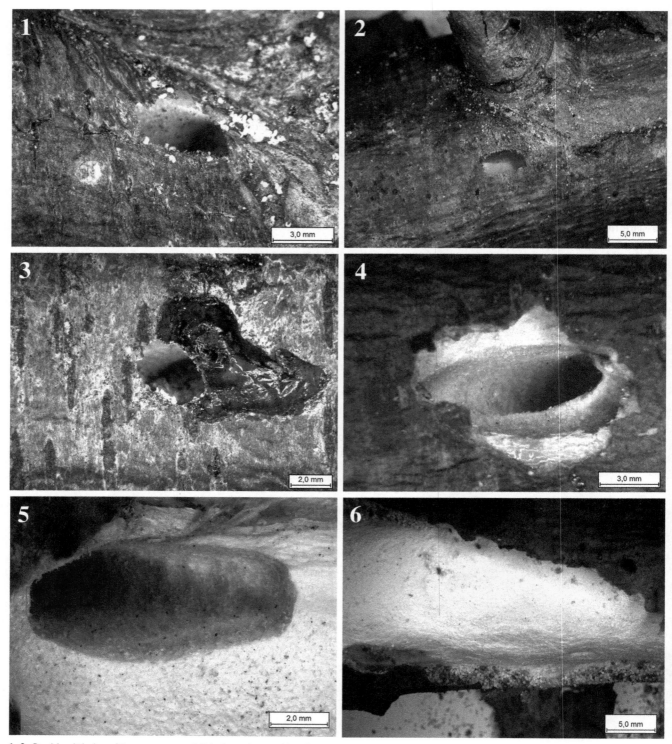

1–3. Ovoid exit holes of *I. tomentosum* visible from the outside of cypress (*Cupressus*) bark
4–5. Exit hole cleared from its bark in the laboratory
6. Mandible traces in the sapwood, detail

The larval galleries of *I. tomentosum* are oval and tend to round off where the imago emerges. The galleries are about 1 cm in diameter, usually 0.6–0.7 cm. This species burrows under the bark and along the sapwood but rarely reaches it. Usually only the outer part of the bark is preserved in the entire cambial part. This gives the thin bark of Cupressaceae the aspect of a sheet of paper on a thick layer of frass. The galleries can be mistaken for those of other fresh-wood capricorn beetles.

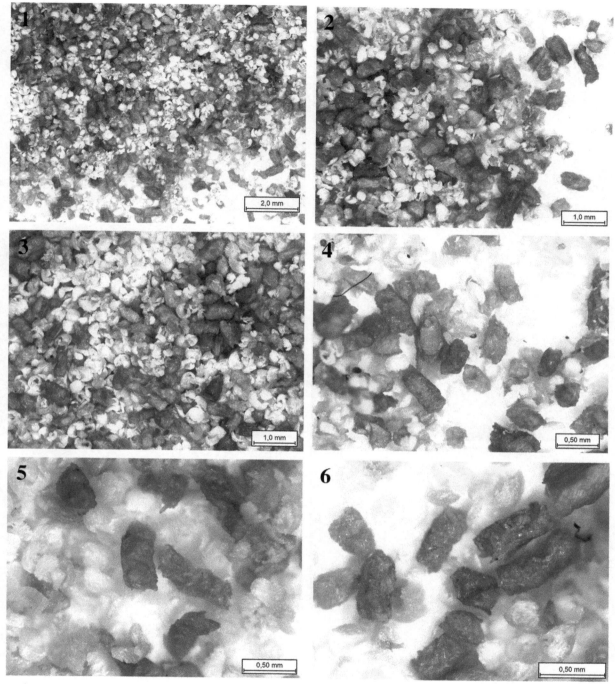

1. Frass of *I tomentosum*, general view
2–3. Heterogeneous frass of *I tomentosum* with wood chips and cylindrical pellets
4–6. Wood chips and cylindrical pellets, detail

The frass is heterogeneous: about 50–60% of it is composed of small faecal pellets called "barrels" (~1 mm long and 0.6 mm wide on average) taking on the colour of the cambium, mixed with small, fairly narrow convex chips (0.3 mm wide) which are more the colour of the sapwood. These chips are produced by the larva's mandibles; they are part of the composition of the visible sawdust expelled from the gallery by the larva. The "barrels" pellets have a more eroded and less smooth surface than those of *Callidium coriaceum*. The wood fibres are not completely fused and digested. Their arrangement gives a slightly twisted aspect to the cylindrical pellet unobserved in the frass of other species currently listed in this book. *I. tomentosum* frass can be mistaken for the frass of other fresh-wood capricorns whose frass is not yet listed. Identification requires to cross-reference all the characteristics of the woody species with those of the frass in SEM.

1. Frass of *Icosium tomentosum* on *Cupressus*, several visible friable cylindrical pellets with wood chips, ×200
2. Undigested convex wood chips, clearly visible wood fibres, detail, ×200
3–4. "Barrel" with agglomerates of various more or less well digested wood fibres, detail, ×350
5. Top of a "barrel" with agglomerates of various wood fibres giving it a twisted aspect, detail, ×500
6. Zoom on an undigested cupressoid punctuation coming from a "barrel", ×3500

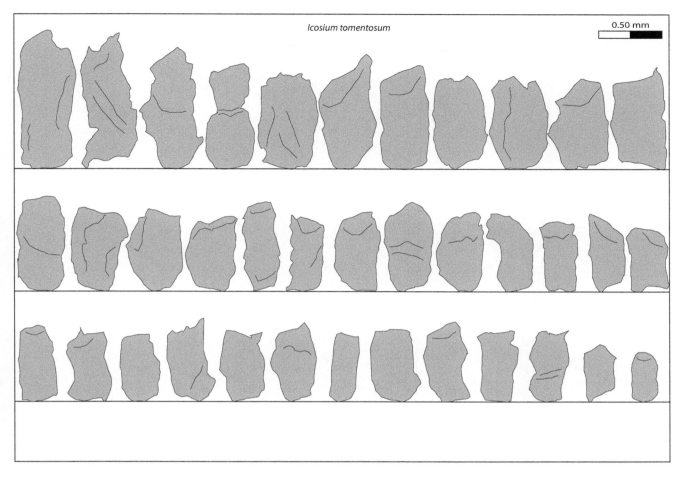

Icosium tomentosum

0.50 mm

Drawings representing 100 or so cylindrical pellets composing the frass of *Icosium tomentosum*. Arranged from the largest to the smallest pellet, harvested from a cypress (*Cupressus*) sample.

Penichroa fasciata (Stephens, 1831)

Capricorn Beetle

Description
This species was once called *Callidium fasciatum*. This rather flattened capricorn is between 6 and 15 mm long. Its antennae are longer than its body. It has more or less dark brown elytra forming a characteristic pattern made of a median band with a light-coloured apical spot.

Distribution
P. fasciata is most often found in the Mediterranean region and more widely in the southern half of France. It is also imported into North America. It is found in the Balkan region except Albania and Macedonia; in the Mediterranean islands, Belgium, France, Italy, Portugal, Spain.

Bioecology
Little information is available on the biology of this species.

Adults are rather nocturnal. They are found on the host plants from the end of June to the end of August.

The eggs are laid in cracks in the wood or in the space between the bark and the sapwood that is created when the wood is dried.

The larvae are very polyphagous and can live in many hardwood and softwood species. They are usually xylophagous, but can adapt to the feeding medium and become polyphagous. In wood, they prefer the sapwood and the subcortical zone.

Cycle of Development
The larval development cycle takes at least 2 years. Pupation takes place in May-June in the wood.

Infested Woods
P. fasciata is found on very dry dead branches and trunks of wood usually stored in the forest after logging.

It prefers deciduous trees such as Judah tree (*Cercis*), pistachio tree (*Pistacia*), fig tree (*Ficus*), oak (*Quercus*), cherry tree (*Cerasus*), broom (*Cytisus*) …

It has been reported on eucalyptus, but also on Aleppo pine (*Pinus halepensis*) or on thuja (*Thuja*).

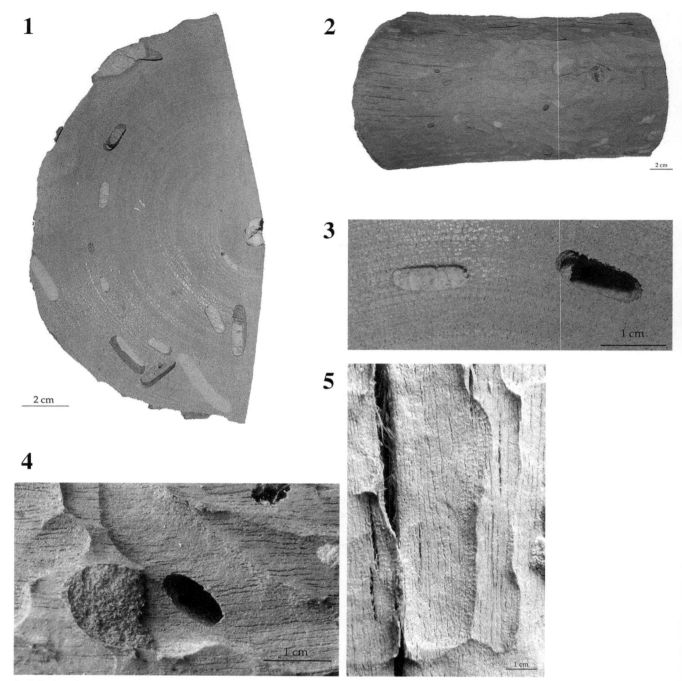

1. Galleries of *P. fasciata* in the sapwood of a fig tree (*Ficus carica*)
2. Subcortical galleries
3. Exit holes, detail
4. Intersecting galleries and frass, detail
5. Mandible tracks in galleries, detail

The oval galleries of *P. fasciata* range from 0.8 to 2 cm in diameter. They are usually found under the bark and can easily reach the sapwood, or even the heartwood if the wood is softwood, preferably in deciduous trees. Infestation is only detectable after the first generation has emerged.

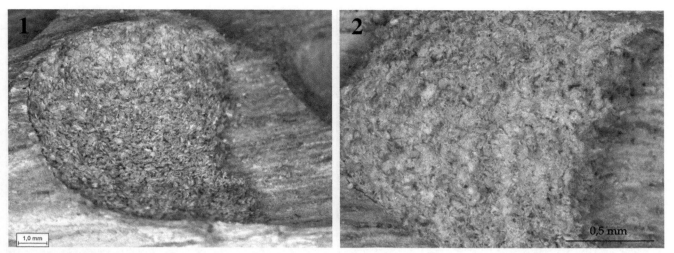

1. Heterogeneous, packed frass of *P. fasciata*
2. Frass composed of many undigested wood fibres (chips), detail

The frass of *P. fasciata* is heterogeneous. It is composed of half of small, friable, cylindrical, easily compactable faecal pellets called "barrels", and half of very fine (<0.3 mm wide) chips (40–50%) that look more like fibres detached from the sapwood of the infested tree. In general, powdery frass is observed, and discernible pellets are rather rare if the wood is old. The frass of *P. fasciata* can be mistaken for the frass of other fresh-wood cerambycids whose frass is not yet listed. Identification requires to cross-reference all the characteristics of the woody species with those of the faecal pellets in SEM.

1. Faecal pellet of *P. fasciata*. Example of crumbling "barrels", packing on top of each other, ×85

2. Compacted frass with many undigested fibres, general aspect, ×130

3–4. "Barrels" with an anarchic arrangement of digested fibres, and presence of poorly digested wood cells, detail, ×110 and ×90

Xylotrechus stebbingi (Gahan, 1906)

Capricorn Beetle

Description
X. stebbingi is 8–21 mm long, brown to reddish-brown, with more or less dense whitish hairs distributed in several horizontal bands on its elytra. Its antennae are shorter than its whole body.

Distribution
Originally from Asia (China and India in particular), this species has settled in the Mediterranean regions (France, Greece, Italy, Kriti, Sardinia, Switzerland), probably following an accidental import in the 1970s.

Bioecology
The biology of *X. stebbingi* is still poorly known due to its recent import. This species seems to adapt to all parts of the tree (bark, sapwood, branch or trunk) and can infest them if the tree is dying on the ground, freshly felled or even stored and therefore relatively dry.

Adults can be observed over a fairly long period from May to September, with a maximum in June and July.

The larva is essentially xylophagous, but it seems to also adapt to the different feeding media it may encounter. The sample presented here comes from a wood storage area awaiting construction in Italy.

Cycle of Development
Its development cycle lasts 1–2 years.

Infested Woods
X. stebbingi mainly infests freshly felled and stored softwood of deciduous trees such as poplar (*Populus*) or alder (*Alnus*), oak (*Quercus*), fig (*Ficus*), ash (*Fraxinus*), hackberry (*Celtis*), mulberry (*Morus*), olive (*Olea*), elm (*Ulmus*), plane (*Platanus*) and sumac (*Rhus*).

2 cm

1. Galleries of *X. stebbingi* on a jujube tree (*Ziziphus spina-cristi*) sample

The galleries of *X. stebbingi* are oval-shaped, like those of all capricorn beetles. They are 0.5–1 cm in diameter. They are located under the bark of the infested tree but also deeper in the wood, especially in the sapwood. The galleries of *X. stebbingi* can be mistaken for those of many cerambycids that infest fresh and stored wood and are not yet listed in this book. Identification is to be refined based on the size of the galleries and the type of infested woody species.

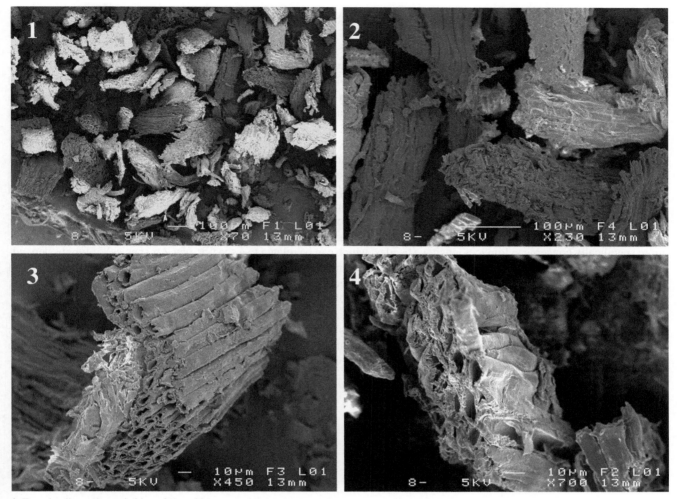

1. Faecal pellet and wood chips of *X. stebbingi* on a jujube tree (*Ziziphus spina-cristi*) sample, ×70
2. Undigested wood fibres, detail, ×230
3–4. Partially digested cells from crumbled barrels, detail, ×450 and ×700

The frass is fine, heterogeneous, and sometimes presents tightly packed cylindrical faecal pellet called "barrels". The pellets disintegrate very easily when touched with the sampling needle. Numerous fine chips (<0.2 mm wide) make up most of the frass (60%). These chips correspond to finely cut wood fibres. The frass of *X. stebbingi* can be mistaken for that of other fresh-wood cerambycids whose frass is not yet described. Identification requires to cross-reference all the characteristics of the woody species with those of the faecal pellets in SEM.

3.1.3   Ciidae

Presentation of the Family

Ciidae form a family of small beetles (0.7–3.5 mm), currently divided into 640 species (Abdullah 1973; Buder et al. 2008). They belong to the super-family Tenebrionoidea and are divided into two subfamilies: Sphindociinae (non-European) and Ciinae. Ciinae are divided into three tribes and 42 genera. In Europe, and more particularly in France, little work has been done on this family (Mellié 1848, Abeille de Perrin 1874, Portevin 1931, etc. for France; Reibnitz 1999 for Europe), which was inventoried and summarised by O. Rose in 2012.

These insects are elongated, cylindrical, often distinctly convex, and display variable punctuations and hairiness. They have short antennae composed of 8–10 segments, not inserted in a dimple. Sexual dimorphism is observed in most species. Both larvae and adults are rather cosmopolitan and fungus-eaters that mainly feed on saproxylic fungi such as polypores (Polyporaceae) or Corticiaceae. They also feed on hyphae and spores under the bark and in decaying wood of trees colonised by these same fungi. They are mainly attracted by the olfactory emissions of their preferred host, mainly at the time of fruiting (Jonsell and Norlander 1995). Healthy fungi are often overlooked: in many cases the beetles are only attracted after fungal spores have been released, often when fruiting is completed and decomposition has started. This course of events more specifically concerns xylomycetophagous beetles, which can be related to a specific type (or species) of fungus depending on the species. Ciidae are important recycling agents in forest ecosystems, many have developed physiological and chemical defence responses to avoid the frequent toxic effects of fungal defence chemicals (Benick 1952; Graves 1960; Lawrence 1973; Fossli and Andersen 1998; Bouget et al. 2005; Orledge and Reynolds 2005; Rose 2012).

This family rather stands apart in this book, but it seemed useful to describe the galleries and wormholes of *Xylographus bostrichoides*. This species was discovered during a diagnosis under a carpenter's bench made of service tree from Mr. Louis Chiorino's collection bought by the Communauté de Communes du Pays des Ecrins located in L'argentière-la-Bessée in the Hautes-Alpes "département". Moreover, the traces it leaves on mushrooms and above all on wood are very characteristic.

Xylographus bostrichoides **(Dufour, 1843)**

Syn: *Cis bostrichoides* Dufour, 1843 = *Xylographus aubei* Mellié, 1848 = *Cis cribratus* Lucas, 1849

Minute Tree-Fungus Beetle

Description

This brown-black insect is approximately 2 mm long and has strongly punctuated elytra with a long, upright, rather loosely packed pubescence. The sutural striae are formed of tightly closed stitches that are erased backwards.

It has protibias whose apex is undulated but strongly denticulated, and comb-like along almost the entire length of their outer edge.

Its antennae are yellow and composed of ten articles, a characteristic which, together with the denticulate protibia on the outer edge, makes it easy to differentiate it from the other genera of this family. Its legs are also yellow.

Confusion remains possible with the genus *Ropalodontus,* but they can be differentiated based on the work of O. Rose (2017).

Distribution

X. bostrichoides is a palearctic species found in Europe (Italy, Germany, Austria, Belarus, Bosnia, Bulgaria, Caucasus, Croatia, Spain, Greece, Hungary, Italy, Poland, the Czech Republic, Romania, Slovakia, Turkey, Ukraine) and North Africa.

Bioecology

It occupies limited spaces but can be abundant (several hundred specimens) in its habitat, including in the mountains.

Adults are rarely seen outside their fungal host, they burrow into the supporting tissues to lay eggs and the entire life cycle is usually short, often ending in 2 months; once the fungus is colonised, many generations may develop, often with the simultaneous presence of larvae, pupae and adults.

Cycle of Development

Around 2 months.

Infested Woods

It is usually found on fungi of the polyporous type, but sometimes tends to use wood adjacent to the fungus.

Its hosts are *Fomes fomentarius, Ganoderma lipsiense, G. lucidum, G. resinaceum, Phellinus tuberculosus.*

It has been reported on sycamore polypore, under poplar bark (*Populus*), hornbeam (*Carpinus*), polypore; beech (*Fagus sylvaticus*) and/or the trunk of Horse chestnut (*Aesculus hippocastanum*).

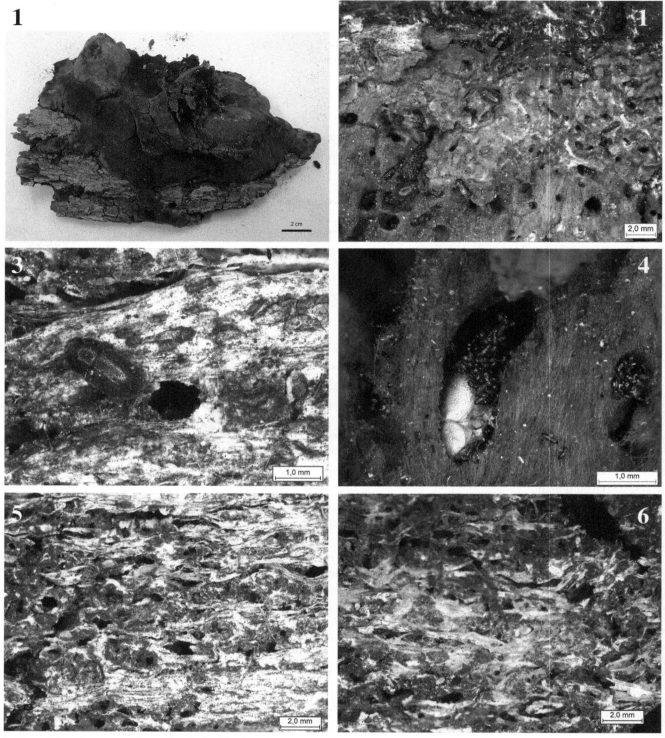

1. View of the reference wood sample colonised by the polypore
2. Numerous galleries and imagoes on the polypore
3. *X. bostrichoides* gallery with an individual adult
4. Individual at the end of pupation in its gallery and with its faecal pellets, in the polypore
5–6. Numerous galleries in the wood, in close proximity to the fungus

The individuals studied here come from a small polypore collected on a decaying deciduous tree. The galleries are irregularly shaped to round, about 1 mm in diameter, and spread in all directions of the fungus and the wood fibres. In the wood, however, the larvae remain in the immediate vicinity of the polypore, feeding only on the wood surface already infested by the fungus (bark, cambium). The larvae do not seem to infest the sapwood. The pupae are mainly found in the fungus.

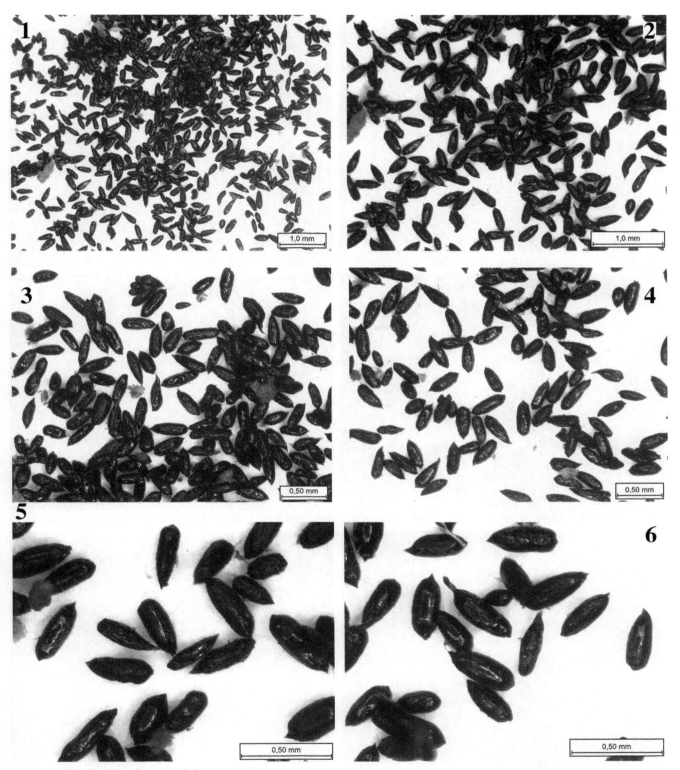

1–4. Frass of *X. bostrichoides*, general view
5–6. Fusiform faecal pellets composing the frass, detail

The frass is homogeneous, and composed of small, smooth, shiny, fusiform faecal pellet. A fusiform pellet of average size 0.39 × 0.14 mm generally has a pointed, sometimes well elongated apex. The pellets are smaller than those produced by *A. punctatum* and larger than those produced by Cossoninae. Under SEM, the pellet is smooth, well digested. A few striae connect the apexes to each other. When the frass is fresh, many small pieces of mycelium are stuck to the surface of the pellet.

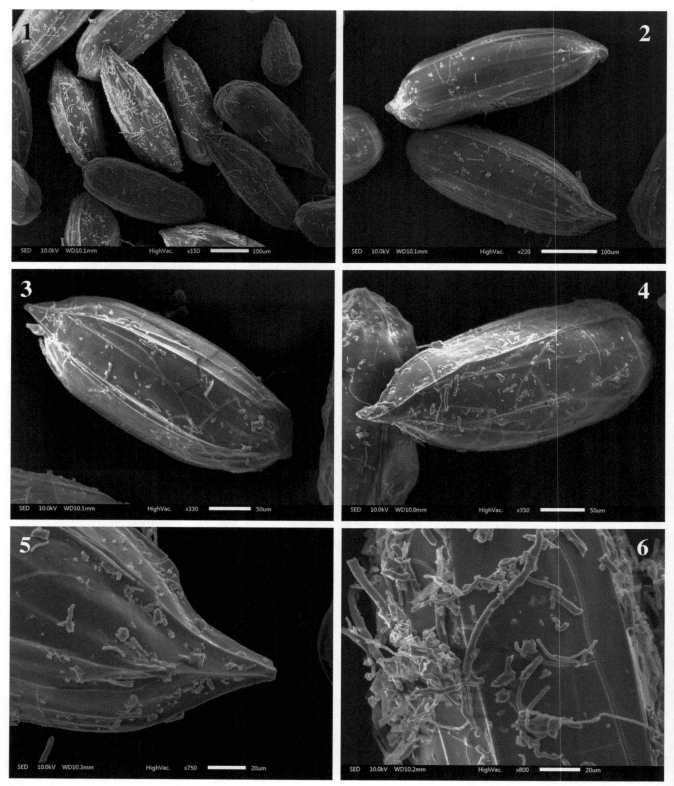

1. Several fusiform faecal pellet of *X. bostrichoides* frass, ×150
2. Two pellets with their apexes and striae, detail, ×220
3–4. One pellet, detail, ×330
5. Pointed, acute apex, detail, ×750
6. Mycelium on faecal pellet, detail, ×800

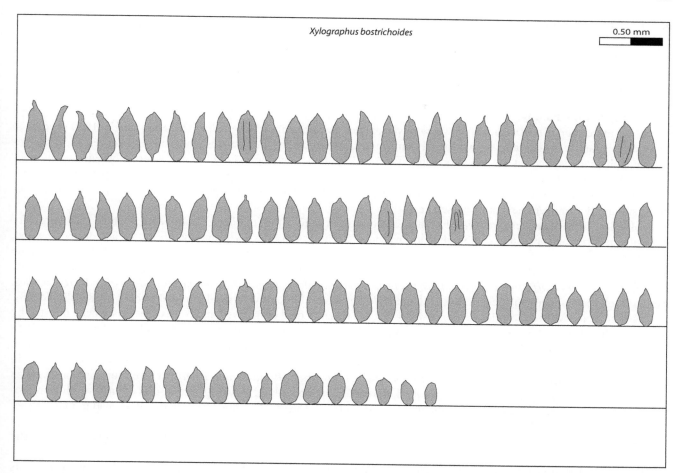

Xylographus bostrichoides

0.50 mm

Drawings representing the fusiform faecal pellets making up the frass of *Xylographus bostrichoides*. Arranged from the largest to the smallest pellet.

3.1.4   Curculionidae

Cossoninae, Molytinae

Wood-Boring Weevils

Presentation of the Family

There are more than 60,000 species of weevils worldwide, described in 4600 genera. There are nearly 1500 species in Europe. This family is generally known for the damage it causes to cereals, herbaceous plants and tree leaves. Most species are phytophagous whether at the larval or adult stage, and feed and develop mainly on angiosperms but also on gymnosperms. However, about 40 species have strictly xylophagous larvae. These species mainly make up the subfamilies Cossoninae and Molytinae, in addition to the genus *Magdalis* in the subfamily Mesoptiliinae, and *Coryssomerus capucinus* in the subfamily Conoderinae, not discussed here.

Scolytidae, or bark beetles, are treated separately because of their different biology and feeding habits: they bore very particular galleries and are rather cambiophagous individuals, whereas wood weevils mainly infest dead and decaying wood. Cossoninae weevil species can infest both deciduous and coniferous trees, with a greater preference for conifers (pine genus): *Amaurorhinus, Aphanommata, Brachytemnus, Choerorhinus, Cossonus, Cotaster, Euophryum, Hexarthrum, Melicius, Mesites, Pentarthrum, Phloeophagus, Pselactus, Pseudophloeophagus, Rhopalomesites, Rhyncolus, Stereocorynes.* Among Molytinae, mainly the genera *Pissodes* and *Hylobius* have a biology very similar to that of bark beetles (Table 3.3).

Cossoninae and Molytinae colonise moist wood that has recently died or decayed and/or has previously been infested by fungi (usually of the polypore type). These two types of infestation usually go together, and accelerate the decomposition process of a wooden item. The genus *Cotaster*, for example, plays an important role in the decomposition of deciduous litter. Sometimes its members are found on driftwood on beaches (*Aphanommata, Mesites*). Finally, in addition to dead wood in the forest, some rather synanthropic species (*Euophryum, Hexarthrum, Pentarthrum, Pselactus*) are found in houses when the construction wood or processed wood (plywood, fibreboard, baseboards …) is wet. These insects often have a high reproductive potential: some species can complete up to 2 full cycles in 1 year. Their presence can therefore lead to major structural collapse of the woods.

Wood weevils are found in most temperate countries. They are small (2–5 mm long) with a rostrum (snout) characteristic of their family.

The usually dark imago has its head extended forward by a rostrum, which bears the antennae and the mouthparts at its tip. In xylophagous individuals, this rostrum allows the female to bore the egg-laying cavity. In general, the larvae are curved and cylindrical, short and fleshy, slightly sclerotic and finely pubescent, and most of them are pale with a darker head.

It is a complex family with descriptions that sometimes overlap from one individual to another. Thus, the identification of certain genera and species requires dissecting the adult individual to observe the genitalia (sexual organs), which are the only organs displaying discriminating characteristics. Moreover, it should be noted that their precise biology is sometimes poorly known, in particular because it is difficult to find certain species and identify them within a regularly updated classification.

Thus, we should be cautious when identifying the galleries and frass of this family. Sometimes it is better not to go beyond the subfamily level because the infestation type and the traces left by the different species are very similar based on the current state of knowledge.

Table 3.3

Species	Sub-family	Root	Branch	Under bark	Bark-less	Trunk	Alive, declining/ dying	Decom-posing	Newly fallen/ felled	Stored	Timber	Decidu-ous	Conifer-ous	Ligneous species infested (non exhaustive list)	European distribution	Cycle	Gallerie diameter
Amaurorhinus (Amaurorhinus) bewickianus (Wollaston, 1860)	Cossoninae	?	x	?	?	x		x		?	x	?	?	Decomposing wood	Balearic Is., France, Greece, Italy, Portugal, Sicily, Spain	?	?
Amaurorhinus (Amaurorhinus) mediterraneus (Folwaczny, 1972)	Cossoninae	?	x	?	?	x		x		?	x	?	?	Decomposing wood	France, Italy, Sicily	?	?
Amaurorhinus (Mazagranus) clermonti (Desbrochers, 1908)	Cossoninae	?	x	?	?	x		x		?	x	?	?	Decomposing wood	France	?	?
Aphanommata filum (Mulsant and Rey, 1858)	Cossoninae		x			x		x		x		?	?	Driftwood on the beach	Albania, Balearic Is., France, Italy, Portugal, Sardinia, Sicily, Spain	?	?
Brachytemnus porcatus (Germar, 1824)	Cossoninae	x	x			x		x				?	?	Decomposing wood	Albania, Azores Is., Balearic Is., Bulgaria, Canary Is., Croatia, Cyprus, Czech Republic, Denmark, France, Germany, Greece, Hungary, Italy, Crete, Latvia, Poland, Portugal, Sardinia, Sicily, Slovakia, Slovenia, Spain, Sweden, The Netherlands, Ukraine	?	?
Choerorhinus squalidus (Fairmaire, 1857)	Cossoninae	x	x			x		x				x		*Alnus, Castanea, Fagus, Fraxinus, Quercus, Populus, Salix, Ulmus*	Bosnia and Herzegovina, France, Italy, Mediterranean is. except Crete and Cyprus; Portugal, Spain	?	?
Cossonus (Caenocossonus) cylindricus (Sahlberg, 1835)	Cossoninae	x	x	x		x		x				x		*Populus, Salix*	Austria, Croatia, Czech Republic, Denmark, Estonia, Finland, France, Germany, Hungary, Italy, Latvia, Poland, Slovakia, Slovenia, Sweden, Switzerland, The Netherlands, Ukraine	?	?

(continued)

Table 3.3 (continued)

Species	Sub-family	Root	Branch	Under bark	Bark-less	Trunk	Alive, declining/dying	Decom-posing	Newly fallen/felled	Stored	Timber	Decidu-ous	Conifer-ous	Ligneous species infested (non exhaustive list)	European distribution	Cycle	Gallerie diameter
Cossonus (Caenocossonus) parallelepipedus (Herbst, 1795)	Cossoninae	x	x	x		x		x				x		Populus, Salix	Austria, Balearic Is., Croatia, Czech Republic, Denmark, Estonia, Finland, France, Germany, Hungary, Italy, Latvia, Norway, Poland, Portugal, Romania, Slovakia, Spain, Sweden, Switzerland, The Netherlands, Ukraine	?	?
Cossonus (Cossonus) linearis (Fabricius, 1775)	Cossoninae	x	x	x		x		x				x		Populus, Salix	Austria, Balearic Is., Bulgaria, Czech Republic, Denmark, France, Germany, Hungary, Italy, Lithuania, Poland, Portugal, Slovakia, Slovenia, Spain, Sweden, Switzerland, The Netherlands, Ukraine	?	?
Cotaster (Cotaster) cuneipennis (Aube, 1850)	Cossoninae		x		x	x		x		x		x		Fagus, Populus, Tillia	Balearic Is., France, Italy, Portugal, Spain	?	?
Cotaster (Cotaster) uncipes (Boheman, 1838)	Cossoninae		x		x	x		x		x		x		Fagus	Austria, Bosnia and Herzegovina, Croatia, France, Germany, Italy, Poland, Slovakia, Slovenia, Switzerland, The Netherlands, Ukraine	?	?
Euophryum confine (Broun, 1880)	Cossoninae	x	x	x	x	x		x		x	x	?	?	Decomposing wood	Austria, Balearic Is., France, Portugal, Spain, the UK	6-12 month	?
Euophryum rufum (Broun, 1880)	Cossoninae	x	x	x	x	x		x		x	x	?	?	Decomposing wood	France, Spain, Switzerland, the UK	6-12 month	?
Hexarthrum capitulum (Wollaston, 1858)	Cossoninae	x	x		?	x		x			x	?	x	Decomposing wood	Austria, Balearic Is., Canary is., Croatia, Czech Republic, France, Greece, Italy, Portugal, Sicily, Spain	?	?

Species	Subfamily										Habitat	Distribution	Development time	Size
Hexarthrum exiguum (**Boheman, 1838**)	Cossoninae	x		x	x	x	x	x	x	x	Decomposing wood, baseboard	**Austria, Balearic Is., Bosnia and Herzegovina, Bulgaria, Croatia, Czech Republic, France, Greece, Hungary, Italy, Poland, Portugal, Slovakia, Spain, Switzerland, The Netherlands, Ukraine**	**Around 8 months**	**<1.5 mm**
Hylobius (Callirus) abietis (Linnaeus, 1758)	Molytinae	x	x	x	x	x	x	x	x	x	Decomposing wood (mainly: *Pinus*)	Austria, Bulgaria, Croatia, Czech Republic, Denmark, Estonia, Finland, France, Germany, Greece, Hungary, Italy, Latvia, Norway, Poland, Portugal, Romania, Sicily, Slovakia, Spain, Sweden, Switzerland, The Netherlands, the UK, Ukraine	Less 1 year	?
Hylobius (Callirus) pinastri (Gyllenhal, 1813)	Molytinae	x	x	x	x	x	x	x	x	x	Decomposing wood (mainly: *Picea, Pinus*)	Austria, Croatia, Czech Republic, Denmark, Estonia, Finland, France, Germany, Hungary, Italy, Latvia, Norway, Poland, Slovakia, Sweden, Switzerland, The Netherlands	Less 1 year	?
Hylobius (Hylobius) excavatus (Laicharting, 1781)	Molytinae	x	x	x	x	x	x	x	x	x	Decomposing wood (mainly: *Larix*)	Austria, Croatia, Czech Republic, Denmark, Estonia, Finland, France, Germany, Hungary, Latvia, Norway, Poland, Romania, Slovakia, Sweden, Switzerland, Ukraine	Less 1 year	?

(continued)

Table 3.3 (continued)

Species	Sub-family	Root	Branch	Under bark	Bark-less	Trunk	Alive, declining/ dying	Decom-posing	Newly fallen/ felled	Stored	Timber	Decidu-ous	Conifer-ous	Ligneous species infested (non exhaustive list)	European distribution	Cycle	Gallerie diameter
Melicius cylindrus (Boheman, 1838)	Cossoninae		x		x	x		x		x	?	?	?	Decomposing wood	Austria, Balearic Is., Bulgaria, Croatia, Czech Republic, France, Greece, Hungary, Italy, Poland, Portugal, Sardinia, Sicily, Slovakia, Spain, Switzerland, The Netherlands, Ukraine	?	?
Melicius gracilis (Rosenhauer, 1856)	Cossoninae		x		x	x		x		x	?	?	?	Decomposing wood	Austria, Balearic Is., Bulgaria, Czech Republic, France, Germany, Hungary, Italy, Poland, Portugal, Sardinia, Sicily, Spain	?	?
Mesites (Mesites) aquitanus (Fairmaire, 1859)	Cossoninae	?	x	x		x		x	?	x		x		*Populus, Salix*	Balearic Is., France, Italy, Portugal, Spain	?	?
Mesites (Mesites) pallidipennis (Boheman, 1838)	Cossoninae	?	x	x		x		x	?	x		x	x	Deciduous and coniferous trees soaked in sea water	Albania, Balearic Is., Croatia, France, Greece, Italy, Portugal, Sardinia, Sicily, Spain	?	?
Pentarthrum huttoni (Wollaston, 1854)	Cossoninae		**x**			**x**		**x**	**x**	**x**		**?**	**?**	**Decomposing wood, baseboard**	**Austria, Belgium, Croatia, France, Germany, Hungary, Ireland, Slovakia, Spain, The Netherlands, the UK**	**7–9 mon-ths**	**<1.5 mm**
Phloeophagus lignarius (Marsham, 1802)	Cossoninae	x	x	?	x	x		x				?	?	Decomposing wood	All Western Europe except Belgium, Croatia, Greece, Ireland, Luxembourg, Macedonia, Sardinia, Sicily	?	?

Taxon									Host	Distribution		Size
Pissodes (Pissodes) castaneus (De Geer, 1775)	Molytinae	x	x	x	x	x	x	x	*Abies, Picea, Pinus*	Austria, Azores Is., Balearic Is., Bulgaria, Canary Is., Croatia, Czech Republic, Denmark, Estonia, Finland, France, Germany, Greece, Hungary, Italy, Latvia, Madeira is., Moldova, Norway, Poland, Portugal, Sardinia, Sicily, Slovakia, Spain, Sweden, Switzerland, The Netherlands, the UK	6 months to 1 year	Around 1 mm
Pissodes (Pissodes) piceae (Illiger, 1807)	Molytinae	x	x	x	x	x	x	x	*Abies, Picea, Pinus*	Austria, Bulgaria, Croatia, Czech Republic, France, Germany, Hungary, Italy, Latvia, Moldova, Poland, Slovakia, Spain, Switzerland, The Netherlands, Ukraine	1–2 years	Around 1 mm
Pissodes (Pissodes) pini (Linnaeus, 1758)	Molytinae	x	x	x				x	*Abies, Picea, Pinus*	Austria, Bulgaria, Croatia, Czech Republic, Denmark, Estonia, Finland, France, Germany, Hungary, Iceland, Italy, Latvia, Norway, Poland, Slovakia, Spain, Sweden, Switzerland, The Netherlands, the UK	6 months to 1 year	Around 1 mm
Pissodes (Pissodes) piniphilus (Herbst, 1797)	Molytinae	x	x	x				x	*Abies, Picea, Pinus*	Austria, Bulgaria, Croatia, Czech Republic, Denmark, Estonia, Finland, France, Germany, Hungary, Italy, Latvia, Poland, Slovakia, Sweden, Switzerland, The Netherlands	6 months to 1 year	Around 1 mm

(continued)

Table 3.3 (continued)

Species	Sub-family	Root	Branch	Under bark	Bark-less	Trunk	Alive, declining/dying	Decomposing	Newly fallen/felled	Stored	Timber	Decidu-ous	Conifer-ous	Ligneous species infested (non exhaustive list)	European distribution	Cycle	Gallerie diameter
Pissodes (Pissodes) validirostris (C. R. Sahlberg, 1834)	Molytinae	x	x	x		x								x	*Abies, Picea, Pinus*	Austria, Bulgaria, Croatia, Czech Republic, Denmark, Estonia, Finland, France, Germany, Hungary, Italy, Latvia, Norway, Poland, Portugal, Slovakia, Sweden, Switzerland, The Netherlands, the UK	6 months to 1 year	Around 1 mm
Pselactus spadix (Herbst, 1795)	Cossoninae	x	x	?		x		x				?	x	Decomposing wood, old wood panelling	Austria, Balearic Is., Bosnia and Herzegovina, Bulgaria, France, Germany, Greece, Hungary, Ireland, Italy, Norway, Poland, Portugal, Sardinia, Sicily, Slovakia, Spain, The Netherlands, the UK	?	?
Pseudophloeophagus aeneopiceus (Boheman, 1845)	Cossoninae	?	?	?		?		x	?	?	?	?	?	Decomposing wood	Austria, Azores Is., Balearic Is., Canary Is., France, Greece, Poland, Portugal, Spain, Switzerland	?	?
Rhopalomesites tardyi (Curtis, 1825)	Cossoninae	?	?	?		?		x	?	?		x		*Ilex aquifolium*		?	?
Rhyncolus (Axenomimetes) reflexus (Boheman, 1838)	Cossoninae	x	x			x		x		x	?	x		*Acer, Aesculus, Alnus, Fagus, Quercus suber, Ulmus*	All Western Europe except Belgium, Balkany region, Greece, Ireland, the UK	?	?
Rhyncolus (Rhyncolus) ater (Linnaeus, 1758)	Cossoninae	x	x			x		x		x	?	x	x	*Castanea, Fagus, Pinus sylvestris, Quercus, Q. suber*	Balearic Is, Bosnia and Herzegovina, Bulgaria, Croatia, Czech Republic, Denmark, Estonia, Finland, France, Greece, Hungary, Italy, Latvia, Norway, Poland, Portugal, Romania, Slovakia, Slovenia, Spain, Sweden, Switzerland, The Netherlands, Ukraine	?	?

Species	Subfamily											Host plants	Distribution		
Rhyncolus (Rhyncolus) elongatus (Gyllenhal, 1827)	Cossoninae	x	x			x	?	x	x		x	*Pinus nigra, P. maritima, P. Halepensis, Abies*	Balearic Is., Croatia, Czech Republic, Denmark, Estonia, Finland, France, Germany, Greece, Hungary, Italy, Latvia, Norway, Poland, Portugal, Romania, Slovakia, Slovenia, Spain, Sweden, Switzerland, The Netherlands, Ukraine	?	?
Rhyncolus (Rhyncolus) punctatulus (Boheman, 1838)	Cossoninae	x	x			x	?	x	x	x	x	*Acer, Aesculus, Alnus, Castanea, Celtis, Fraxinus, Populus, Quercus, Ulmus*	Austria, Balearic Is., Bosnia and Herzegovina, Czech Republic, France, Germany, Hungary, Italy, Poland, Portugal, Sardinia, Sicily, Slovakia, Spain, Sweden, Switzerland, The Netherlands, Ukraine	?	?
Rhyncolus (Rhyncolus) sculpturatus (Waltl, 1839)	Cossoninae	x	x			x	?	x	x		x	*Picea*	Austria, Balearic Is., Croatia, Czech Republic, Denmark, Estonia, Finland, France, Germany, Greece, Italy, Latvia, Norway, Poland, Slovakia, Spain, Sweden, The Netherlands	?	?
Rhyncolus (Rhyncolus) strangulatus (Perris, 1852)	Cossoninae	x	x			x	?	x	x		x	*Pinus nigra, P. maritima, P. Halepensis*	Balearic Is., Croatia, France, Italy, Portugal, Spain	?	?
Stereocorynes truncorum (Germar, 1824)	Cossoninae	?	x			x	?	x	x	?	x	Mainly coniferous Reported on *Fraxinus, Quercus, Salix*	Austria, Balearic Is., Belgium, Bosnia and Herzegovina, Czech Republic; Denmark, Estonia, France, Germany, Greece, Hungary, Italy, Latvia, Poland, Portugal, Romania, Slovakia, Spain, Sweden, Switzerland, The Netherlands, the UK	?	?

Hexarthrum exiguum (Boheman, 1838)

Wood-Boring Weevil

Description

Of variable length (between 2 and 5 mm), *H. exiguum* is light to dark brown depending on the type of feeding support. Its rostrum is prominent and broad. Its antennae are bent and inserted halfway along the rostrum.

The individuals currently being bred for this work at the CICRP evolve on base-boards made of composite deciduous wood.

Distribution

H. exiguum is commonly found in temperate countries and regions: Austria, the Balearic Islands, Bosnia and Herzegovina, Bulgaria, Croatia, the Czech Republic, France, Greece, Hungary, Italy, Poland, Portugal, Slovakia, Spain, Switzerland, The Netherlands, Ukraine.

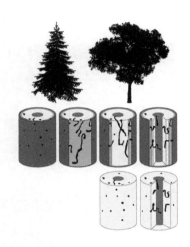

Bioecology

This species is found throughout the year on decaying wood (especially in forests) or on very damp timber (flooring, baseboards ...).

The eggs (20–40) are laid in the cracks and irregularities of the wood. The larva is white and strongly arched. It generally bores galleries in the same direction as the wood grain for 6 months to 1 year. It rather infests the sapwood, but can also infest the heart-wood or even healthy parts of the wood after or at the same time as the lignivorous fungus.

Adults live longer than most insects, about 16 months, and also feed on wood.

Cycle of Development

The development cycle of the larva is very short compared to that of most other xyloph-agous insects. There can be 1 cycle every 8 months or less if environmental conditions (high temperature and humidity) are favourable.

Infested Woods

This species seems to infest both deciduous and coniferous trees indiscriminately, as the driving parameters of its colonisation process are the general state of the structure and the thermo-hygrometric conditions.

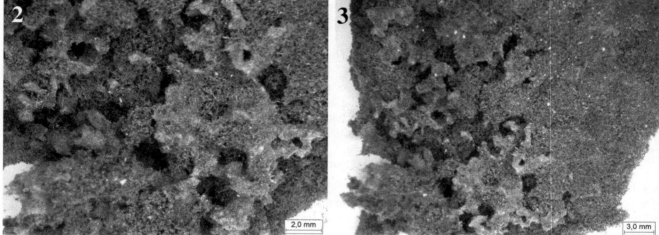

1. Breeding of *Hexarthrum exiguum* on a baseboard, untouched painted surface (but high infestation below)
2–3. Many intersecting galleries, very strong infestation and degradation of the object, detail

The galleries are irregular to round. The adult exit holes vary from 1 to 2 mm in diameter, or even less in the galleries of young larvae. The wood can also be eaten by adult individuals.

The larval galleries run through the sapwood and the heartwood alike and do not necessarily follow the direction of the wood grain. Similarly, on processed wood such as plywood or compressed wood baseboards made from several tree species, the larvae burrow in all directions of the object, preferably in the opposite direction of or avoiding a possible painted surface. The galleries of a whole colony can altogether look like intersecting networks.

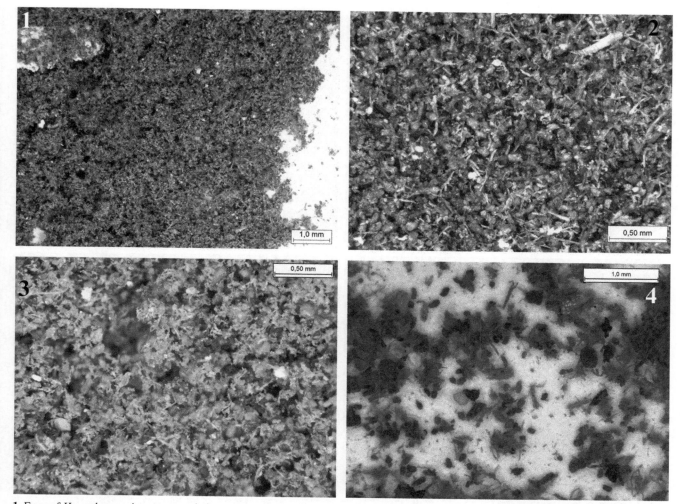

1. Frass of *Hexarthrum exiguum,* general view
2–4. Heterogeneous frass with little fusiform pellets to be differentiated from partially digested or undigested wood fibres and other small wood scraps

The whole frass is heterogeneous. It is composed of two assemblages in approximately the same proportions: on the one hand, fine poorly digested or even undigested wood scraps, mainly composed of fibres, and on the other hand, small fusiform faecal pellets with rounded ends or sometimes with a pointed apex. The pellet generally does not exceed 0.2 mm in length and tends to crumble because it is composed of sometimes poorly digested agglomerates of wood fibres (some wood cells are preserved).

The frass of *Hexarthrum exiguum* can be mistaken for that of other Cossoninae. The difference with *Pentarthrum huttoni* frass lies in the different proportions of pellets/scraps and in the heterogeneity of the frass as a whole, but possible confusion with the frass of other so far unlisted species should not be excluded.

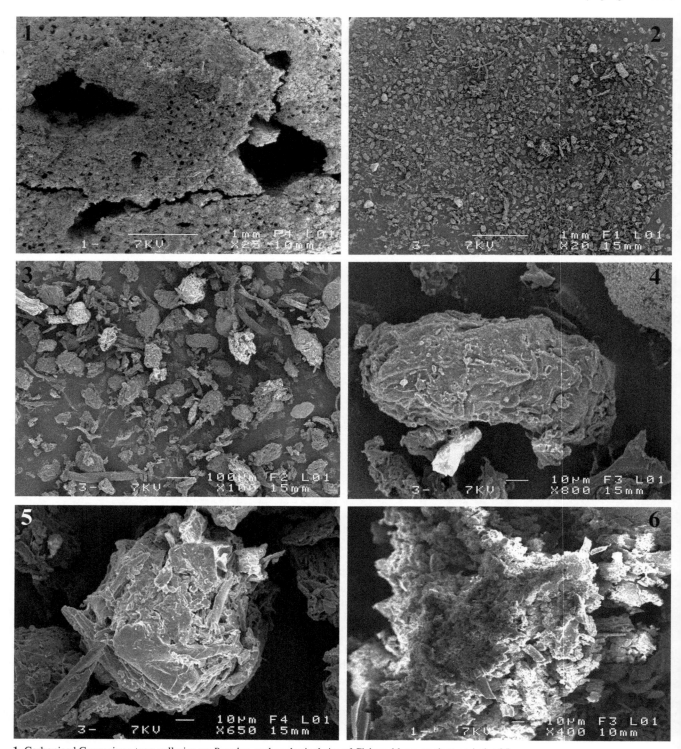

1. Carbonised Cossoninae-type galleries on *Populus*, archaeological site of Châteaubleau, ancient period, ×25
2. Frass of *H. exiguum* on contemporary baseboard, several irregular pellets visible, ×20
3. Faecal pellet and wood chips, detail, ×100
4. One pellet, detail: agglomerate of various visible wood fibres, ×800
5. Zoom on the apex of a pellet. Various arrangements of poorly digested wood fibres, ×650
6. Miscellaneous partially digested wood scraps, detail, ×400

Hexarthrum exiguum

0.50 mm

Drawings representing 100 or so irregular fusiform faecal pellets composing the frass of *Hexarthum exiguum*. Arranged from the largest to the smallest pellet, collected from the composite wood sample mentioned above.

Pentarthrum huttoni (Wollaston, 1854)

Wood-Boring Weevil

Description

P. huttoni is 2.7–4 mm long and varies in colour from blackish brown to reddish brown. Its thorax is darker than its elytra.

The adult is subdepressed with shiny, hairless cylindrical elytra, and has iron-coloured legs and antennae. The antennal funiculus consists of five articles.

Elderly larvae are 3.5 mm long and 1 mm wide.

P. huttoni can be mistaken for *Euophyum confine*, a morphologically and biologically very similar species; *E. confine* differs from *P. huttoni* by a strong constriction of its head, behind its eyes, and the absence of a strong elevation of the ninth interstripe of the apex of its abdomen.

Distribution

Probably native from South America, this insect has spread throughout Europe as well as North America and Canada.

It is a synanthropic species, and although its range is Western Palearctic, its occurrence is rare, sporadic and generally very localised.

It is found in the following countries: Austria, Belgium, Croatia, France, Germany, Hungary, Ireland, Slovakia, Spain, The Netherlands, the United Kingdom.

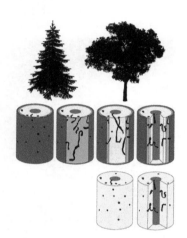

Bioecology

Adults are found year-round and their presence can be spotted from dead specimens at windows and other light sources because they are attracted by light during dispersal.

They usually stay close to the material they infest. When thermo-hygrometric conditions are favourable, they can reproduce continuously. This species is associated with decaying or highly humid wood.

One week after mating, females lay their eggs singly in cracks in the wood, which must be moist because this insect does not develop or lay its eggs in dry wood. If necessary, the female can drill into the wood with its rostrum to lay an egg. The female is not very prolific: it lays 20–30 eggs over a period of 3 months. The larvae emerge after 2–3 weeks.

The larvae usually tunnel parallel to the surface of the wood, but can also bore random galleries by following the moisture gradient of the wood. They build a pupal chamber lined with fungal hyphae just below the surface of the wood. They can digest cellulose and hemicellulose, but lignin is excreted in the faeces.

Cycle of Development

Larval development consists of five stages and lasts 6–8 months. The pupal stage lasts 2–3 weeks. Adults can live up to 16 months; they feed on the wood and also participate in its degradation. The complete cycle from egg to adult lasts 7–9 months.

Infested Woods

This insect can degrade both the sapwood and heartwood of hardwoods and softwoods when the wood is highly moistened and pre-digested by lignivorous fungi. It is present in wet wood in houses, cellars and sheds, etc.

It infests the same wood species as *Hexarthrum exiguum* does.

1. View of *Pentarthrum huttoni* breeding on wet coniferous manufactured objects
2–3. Longitudinally visible galleries with frass
4. Irregular galleries, detail

The galleries are irregular to round. The emergence holes measure 1–2 mm and are irregularly shaped. The wood can also be gnawed by adult individuals. The larval galleries indifferently run through the sapwood or the heartwood and do not necessarily follow the direction of the wood grain. Similarly, on processed wood such as plywood or compressed wood baseboards made from several wood species, the larvae burrow in all directions of the object, preferably in the opposite direction of or avoiding a possible painted surface. The galleries of a whole colony can altogether look like intersecting networks.

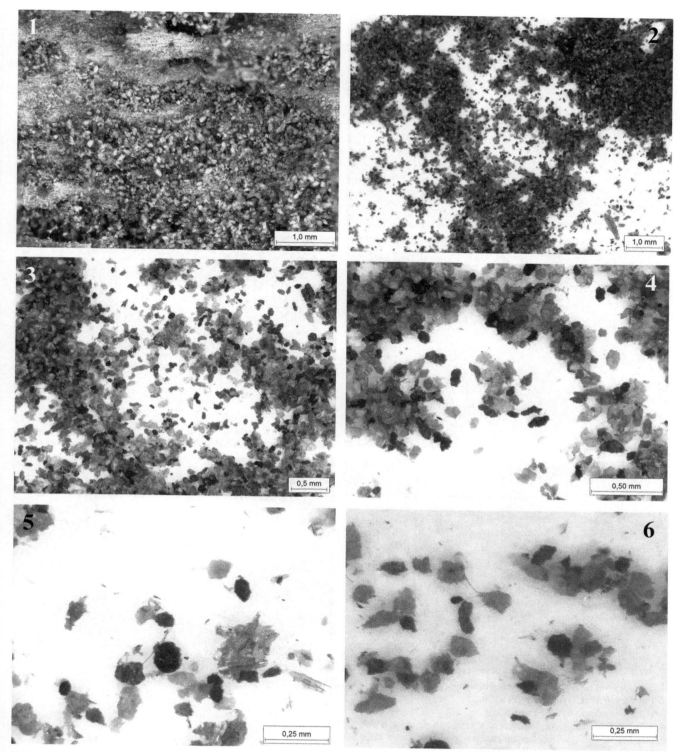

1. Frass of *Pentarthrum huttoni* filling the gallery
2. Heterogeneous frass, general aspect
3–6. Fusiform pellets and wood scraps composing the frass, detail

The frass as a whole is heterogeneous. About 20–30% of it is composed of fine, poorly or even undigested wood scraps, mainly composed of fibres, and 70–80% of it is made up of small fusiform faecal pellets with rounded ends or sometimes with a pointed apex. The pellet generally does not exceed 0.2 mm in length and tends to crumble because it is composed of agglomerates of sometimes poorly digested wood fibres (some wood cells are preserved).

P. huttoni can be mistaken for other Cossoninae. The difference with *Hexarthrum exiguum* lies in the different proportions of the heterogeneity of the frass as a whole, but possible confusion with other so far unlisted species should not be excluded.

1. Heterogeneous frass of *P. huttoni* on contemporary manufactured wood, several visible irregular pellets and some scraps, ×200
2. Fusiform pellets varying in size, ×250
3. Pellets and wood scraps, detail, ×350
4. A few pellets, detail. Agglomerate of various visible wood fibres, ×500
5. One pellet, detail. Various arrangements of poorly digested wood fibres, ×1000
6. Partially digested woody cell, ×2500

Pentarthrum huttoni

0.50 mm

Drawings representing 100 or so irregular fusiform faecal pellets composing the frass of *Pentarthrum huttoni*. Arranged from the largest to the smallest pellet, collected from the composite wood sample mentioned above.

Scolytinae

Bark Beetles

Presentation of the Family

This subfamily of Curculionidae corresponds more precisely to the so-called cambiophagous insects. During their larval development, these pests consume the cambium located between the sapwood and the bark of living or dying trees. This type of infestation makes bark beetles dangerous forest pests with a high environmental and economic impact. Their activity, as well as inoculation/symbiosis with a lignivorous fungus during egg-laying that blocks the natural defence responses of the infested tree, usually leads to the death of the tree or even of the entire woody population of a sector (balsam fir, elm grove, etc.).

The group includes about 6000 species divided into 230 genera and 28 tribes round the world. There are 1000 species for the whole palearctic region, about 100 species in Central Europe, 140 species in France and 66 in the United Kingdom. A large number of conifers and deciduous trees are home to these species whose larvae cause significant damage when their galleries are bored across vast areas of bark, destroying the phloem vessels. Mycophagous bark beetles that infest the sapwood are very rare. The different species are obligate pests of either hardwoods or softwoods, but very rarely of both at the same time. Softwoods are the most impacted. Some rare bark beetle species have been found on herbaceous plants (*Thamnurgus*) or in seeds (*Coccotrypes*).

Certain bark beetle species prefer certain parts of the tree: large branches, twigs; bark of different thicknesses (*Pityokteines* under thin bark); base of the trunk and roots (*Hylastes ater*, *Dendroctonus micans*); branches and tree top (*Ips*) …

Their development and the number of annual generations is highly dependent on climate and temperature. Activity can start as soon as the temperature reaches 5 °C; normal activity occurs between 10 and 15 °C, swarming between 16 and 18 °C, optimal activity between 18 and 29 °C and hyperactivity between 30 and 40 °C.

Scolytinae—e.g., Platypodinae—are not dealt with in the present atlas because they are rather tropical, and are different from other Curculionidae: they are small beetles, 1–9 mm but most often 1–4 mm in length, cylindrical or slightly flattened on the back, and most of them are strongly sclerosed. They are fairly good fliers, and can be distinguished from the other groups because they present pre-gular sutures, their rostrum is short or absent, and their anterior tibias have teeth or bristles on their outer edge.

Oviposition is highly variable from one species to another, from 10–12 to 300 eggs.

The white, curved, apodic larvae of most species are xylophagous. They bore their galleries starting from the gallery made by the adults (oviposition gallery or maternal gallery). Some bark beetles are dependent on symbiotic fungi (e.g., *Ambrosiaemyces zeylanicus*, formerly *Ambrosia*, or the ascomycete *Pezizomycotina*) with which they seed their galleries, and consume once the wood has been pre-digested. The bark beetle cannot directly digest the wood because of the presence of lignin and cellulose. The larva does not ingest the pre-digested plant fibres, it rather ingests the fungus itself.

It is possible to distinguish various groups within this family:

- Strictly primary pests attack vigorous trees.
- Some secondary pests become primary depending on certain climatic contexts and during outbreaks (*Ips typographus*, *Tomicus piniperda, Dendroctonus micans* …) (Dajoz 1980).
- Some harmful secondary pests attack weakened trees (due to tree overpopulation, drought spells, damage caused by defoliating caterpillars …) and transmit cryptogamic diseases (Dutch elm disease caused by *Scolytus multistriatus*) to them (*Ips, Pityokteines, Orthotomicus, Pityophtorus, Hylesinus, Leperesinus* …).
- Some secondary pests again infest trees that have already been felled or remain standing but heavily affected.
- Some are pests of more fragmented, decaying wood (*Dryocoetes, Hylurgops*).

Their cycle consists of several stages:

1. Swarming: the bark beetle cycle begins with the choice of the host tree to be colonised. This is a testing process carried out by various individuals known as "pioneer bark beetles". Once the host has been chosen after being tasted, the pioneers send out aggregation pheromones (different for each species) that attract other congeners, both male and female.
2. Aggregation corresponds to the infestation and distribution of bark beetles on several hosts, forming a "focus".
3. Colonisation involves boring into the cambium of the tree, and then the maternal gallery forms a network characteristic of the species, along with the larval galleries. By inoculating a fungus during colonisation, this family is the vector of a tree disease called graphiosis.
4. The epidemic or even booming of the population *per* cubic metre of dead wood, felled after a storm or sick and still standing, can be huge. The limits are the availability of host trees, and the degree of competition between the scolytes and the

predators of larvae. Some species can swarm and cause considerable economic damage. This was the case in France, for example, after the storm of 1999 when 5 million m³ of wood were infested.

The galleries of Scolytinae appear to have a characteristic shape for each species (Balachowsky 1949, 1963; Nageleisen et al. 2010). Large-scale studies and observations would need to be carried out to provide more information about this point. A gallery system consists of three elements: the entrance hole (bored by the male or the female); the egg-laying gallery or galleries (or maternal galleries), sometimes with mating chambers, drilled by the adults and kept clean of wood waste or scraps, their length varies according to the species; and the larval galleries (containing the frass) bored by the larvae, all starting from the central maternal gallery and ending in a more or less widened cul-de-sac where pupation takes place. The young imago then punctures an exit hole through the protective and nourishing bark.

There are about ten different types of maternal galleries: single, double or triple (vestibules or bridal chamber) which can be longitudinal (in the same direction as the wood fibres) or across (perpendicular to the fibres); irregular, curved galleries; galleries that overlap in single (two galleries) or double (four galleries) (forming an H), the bridal chamber is then slightly apart; star-shaped galleries (with the mating chamber in the centre) following the direction of the fibres or not; finally, the so-called false star-shaped galleries, which are subcircular, because they are in fact a family chamber where the eggs are deposited all around the outer walls. The larval galleries forming the star radiate in all directions.

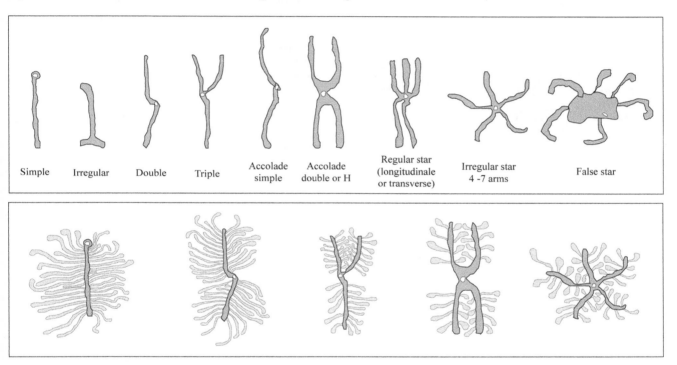

| Simple | Irregular | Double | Triple | Accolade simple | Accolade double or H | Regular star (longitudinale or transverse) | Irregular star 4 -7 arms | False star |

Drawings below based on Balachowsky (1949), Dajoz (1980), Nageleisen et al. (2010). (**a**) Maternal galleries only. (**b**) Maternal galleries and larvae, examples of some cases. From left to right, e.g., *Scolytus scolytus* (on *Ulmus*), *Ips typographus* (on *Picea*), *Ips sexdentatus* (on *Pinus*), *Pityokteines curvidens* (on *Abies*), *Pityogenes chalcographus* (on *Picea*) (Table 3.4).

Table 3.4

Species	Deciduous	Coniferous	Ligneous species	Maternal gallery	Direction in relation to the fibres
Cryphalus piceae		x	*Abies, Picea*	False star	Various
Dendroctonus micans		x	*Picea*	False star	Various
Hylesinus crenatus	x		*Fraxinus*	Simple or accolade	Perpendicular
Ips acuminatus		x	*Pinus* Mainly *sylvestris*	Star	Various
Ips amitinus		x	*Picea*	Double or double accolade (H)	Longitudinal
Ips cembrae		x	*Larix*	Double	Longitudinal
Ips duplicatus		x	*Picea*	Triple	Longitudinal
Ips sexdentatus		x	*Pinus*	Double	Longitudinal
Ips typographus		x	*Picea*	Double, triple	Longitudinal
Leperesinus fraxini	x		*Fraxinus*	Accolade	Perpendicular
Scolytus intricatus	x		*Quercus*	Simple	Perpendicular
Scolytus multistriatus	x		*Ulmus*	Simple	Longitudinal
Scolytus ratzeburgi	x		*Betula*	Simple	Longitudinal
Scolytus rugulosus	x		Fruit trees	Simple	Longitudinal
Scolytus scolytus	x		*Ulmus*	Simple	Longitudinal
Pityogenes chalcographus		x	*Picea*	Star	Various
Pityokteines curvidens		x	*Abies*	Double accolade (H)	Perpendicular
Pityophthorus pityographus		x	*Picea*	Double	Perpendicular
Pteleobius kraatzi	x		*Ulmus*	Simple	Perpendicular
Taphrorychus bicolor	x		*Fagus*	Star	Various
Tomicus piniperda		x	*Pinus*	Simple	Longitudinal

Ips typographus (Linnaeus, 1758)

Syn: *"Bostrychus" typographus*
Engraver Beetle, European Spruce Bark Beetle

Description

Ips typographus is 4–5 mm in length. It is light to dark brown depending on the type of feeding medium and is covered with light-coloured hairs. Its elytra present a declivity, with four teeth at their tip.

I. typographus can be mistaken for *Ips sexdentatus,* which is longer (5–8 mm).

Distribution

Ips typographus can be found mainly in the mountains (Alps and Jura) and high plains. It is very common in Switzerland but spreads more generally throughout continental Europe in spruce areas.

Bioecology

Adult swarming can take place over several kilometres as early as spring. For this to happen, a few successive rainless days at 18–20 °C, frostless at night, are required. They colonise weakened trees first. Swarming can occur late (April to June) as long as the temperature remains around 16–18 °C.

The male initiates the egg-laying process. One or two females join him. After fecundation, the females bore egg-laying (or maternal) galleries parallel to the wood fibres in the cambium. They lay the eggs (20–50) regularly in egg-laying notches on either side of the gallery. The female can have two different egg-laying spells, forming sister generations, with a double or triple maternal gallery.

As soon as the larvae hatch, they bore a sinuous gallery perpendicular to the maternal gallery, starting from the egg notch. The larval gallery gradually widens and ends in the pupal "cradle". Depending on the stage reached in winter, the larva, the pupa or the imago overwinters under the bark of the host tree, or even in the soil for mature adults.

Subcortical larval development and the presence of fungi carried by the insect during the drilling operation lead to the destruction of the sap-conducting tissues, causing the bark to fall and thus leading to the rapid death of the host. The death of the infested tree occurs within weeks or months after the end of bark beetle development.

The fungi carried by *I. typographus* are blue-stain agents (*Ophiostoma* spp.) which contribute to decreasing the economic value of the colonised wood.

Cycle of Development

Development is highly temperature-dependent. There are two complete generations in the year, from April to October, three in exceptionally warm years and only one at high altitudes.

Infested Woods

I. typographus is a parasite of weakened trees that mainly infests spruce (*Picea* spp.), especially trunks and branches. It can infest trees more than 20 cm in diameter. It is occasionally found on pine (*Pinus*), fir (*Abies*) and larch (*Larix*). It preferably infests freshly felled, weakened or mutilated trees. If the infestation becomes epidemic, the insects may then infest increasingly healthy wood or even trees in full vigour, and thus become a primary pest.

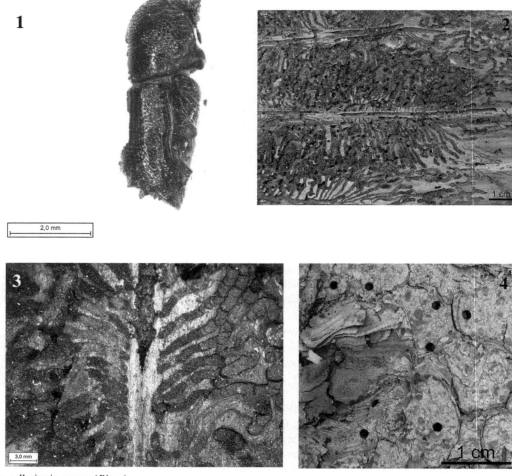

1. Imago found in galleries in spruce (*Picea*)
2. Double mother galleries and larval galleries visible under bark
3. Egg-laying galleries and larvae, detail
4. Circular exit holes

Maternal and larval galleries are irregularly shaped and circular to well rounded. This shape seems to depend on the thickness and hardness of the bark. The galleries are 1.5–2.5 mm in diameter.

The maternal gallery is double or triple, free of excrements and in the same direction as the wood fibres (longitudinal). It spreads out under the bark, in the cambium of the tree along the sapwood but without actually reaching it, over a length of 5–10 cm, sometimes more.

The larval galleries start at regular intervals from the egg-laying gallery. Then, each gallery (one gallery = one individual) is bored along the cambium over a few centimetres (generally 4–5) to end in a pupation chamber that is slightly wider than the rest of the gallery. The adult will burrow through the bark from its chamber to emerge.

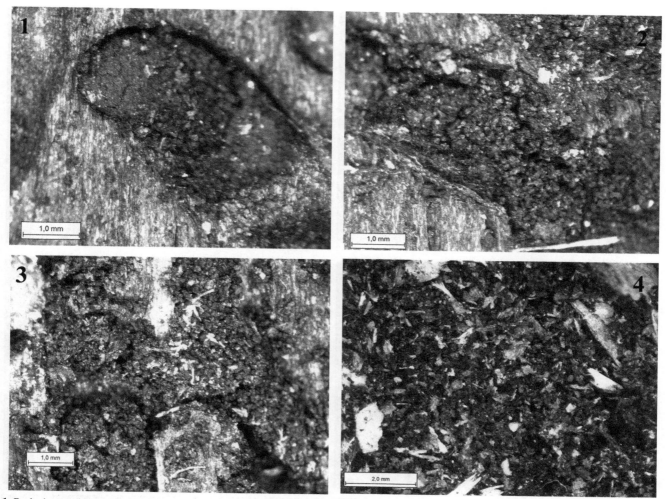

1. Packed, granular, heterogeneous frass of *I. typographus* in a pupal chamber
2. Heterogeneous granular frass in a larval gallery
3–4. Heterogeneous frass composed of small fusiform pellets and partially digested scraps, detail

The frass is heterogeneous. It is composed of two assemblages containing approximately the same proportions of fine, undigested or poorly digested (fibres) wood scraps on the one hand, and small fusiform faecal pellets with rounded ends or sometimes with a pointed apex on the other hand. One pellet is usually no longer than 0.4 mm. SEM observations can show fungal hyphae rolling up the faecal pellet.

The frass is very similar to that of other Scolytinae. It can be mistaken for the frass of Cossoninae/Molytinae, which is also heterogeneous. Differentiation is made based on the typical galleries of Scolytinae and their localisation in the wood. Finally, the proportion of frass heterogeneity and the presence of fungal hyphae wrapped around the pellets can help in the identification.

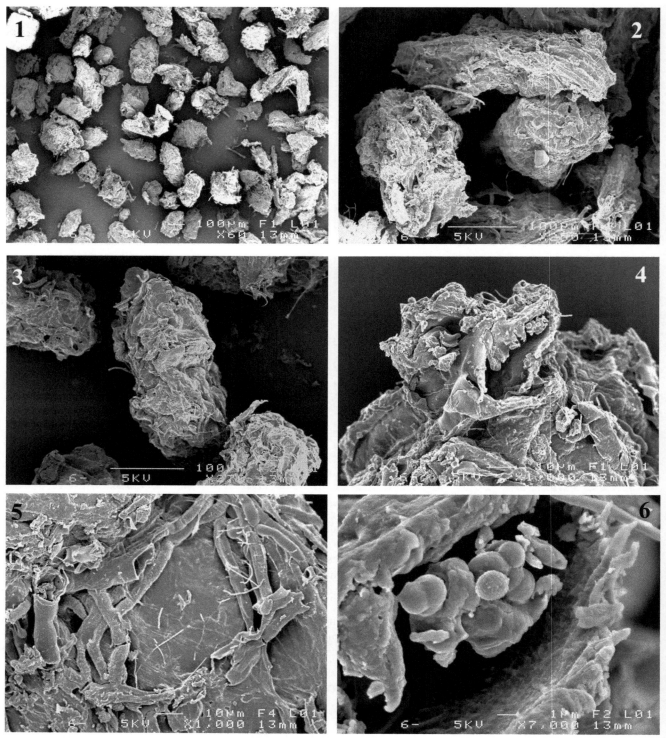

1. Fusiform, heterogeneous faecal pellets of *Ips typographus* on spruce bark, ×60
2. Fusiform pellets, detail, ×250
3. Fusiform pellet with arrangement of wood and cambium fibres, detail, ×270
4. Acute apex of a fusiform pellet, detail, ×1000
5. Zoom on the arrangement of wood and cambium fibres composing a fusiform, half-digested cellular pellet, ×1000
6. Zoom on fungi present in the remains of poorly digested wood fibres of spindle-shaped pellets, ×7000

Scolytus multistriatus (**Marsham, 1802**)

Little Elm Bark Beetle

Description
S. multistriatus is a relatively small bark beetle, 2–3.5 mm in length. It has a big tuber (protuberance) located on the posterior end of its second abdominal sternite. It is anatomically close to *Scolytus scolytus*, the great elm bark beetle.

Distribution
This species is widely distributed throughout Europe as well as in Western Asia, North Africa and some states of the United States.

It proliferates in areas where elm (*Ulmus* spp.) grows.

Bioecology
Its biology is very close to that of its congener *Scolytus scolytus*.

The smell of the host tree plays an important role in its attractiveness. Adults can be seen from May to October and prefer the tree tops. Individuals of this species fly in the spring and infest dying elms. The small elm bark beetle produces an aggregation pheromone to attract its congeners. The larvae develop during the summer months from a single maternal gallery. They overwinter as larvae in trees, and mature in the winter or spring following colonisation.

Like the other Scolytinae, *S. multistriatus* is a cambiophagous species that burrows galleries under the bark of old, weakened or freshly felled living wood.

Together with the fungus *Ophiostoma ulmi*, of which it is the vector, it causes elm disease, i.e., dieback of branches, rapid yellowing and drying of leaves, and accelerated decline of the tree.

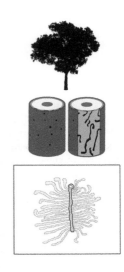

Cycle of Development
The little elm bark beetle has a variable life cycle depending on latitude, with two (or even three) generations in the southern part of its range. More generally, its life cycle lasts 1 year.

Infested Woods
It is almost an obligate parasite of elm (*Ulmus*), but can also be found on poplar (*Populus*) and oak (*Quercus*).

1. Sample of elm (*Ulmus minor*) with *S. multistriatus*-infested subcortical zone, which can be distinguished from the three cerambycid galleries in the same wood based on the size of the galleries and the geometric pattern of bark beetle galleries
2. Two simple, longitudinal egg-laying galleries with larval galleries, overlapping in places, detail

Maternal and larval galleries are irregularly shaped and circular to well rounded. Their diameter varies between 1 and 2 mm. The maternal gallery is simple, frass-less, and follows the direction of the wood fibres (longitudinal). It spreads out under the bark, in the cambium of the tree along the sapwood but without actually reaching it, over a length of 3–5 cm, sometimes more.

The larval galleries start at regular intervals from the egg-laying gallery. Then, each gallery (one gallery = one individual) is bored along the cambium over a few centimetres (generally 3 cm) to end in a pupation chamber, which is not always clearly visible because it is slightly wider as compared to the rest of the gallery. The adult will bore through the bark from its chamber to emerge.

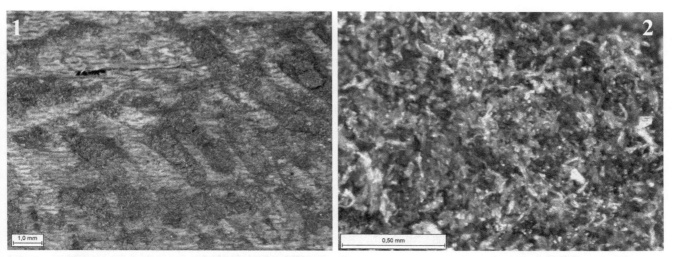

1–2. Heterogeneous, granular frass packed in larval galleries in *Ulmus minor*

The frass is heterogeneous. It is composed of two assemblages in approximately the same proportions: fine, undigested or poorly digested (fibres) wood scraps (sometimes in a greater proportion: 60%) on the one hand, and small fusiform to sub-spherical pellets with rounded ends or sometimes with a pointed apex on the other hand. These pellets are very friable. One pellet does not generally exceed 0.3 mm in length.

The frass from the reference sample is very compacted because it is old.

The frass is very similar to that of other Scolytinae. Therefore, one should mind possible confusion with the frass of Cossoninae/Molytinae frass, which is also heterogeneous.

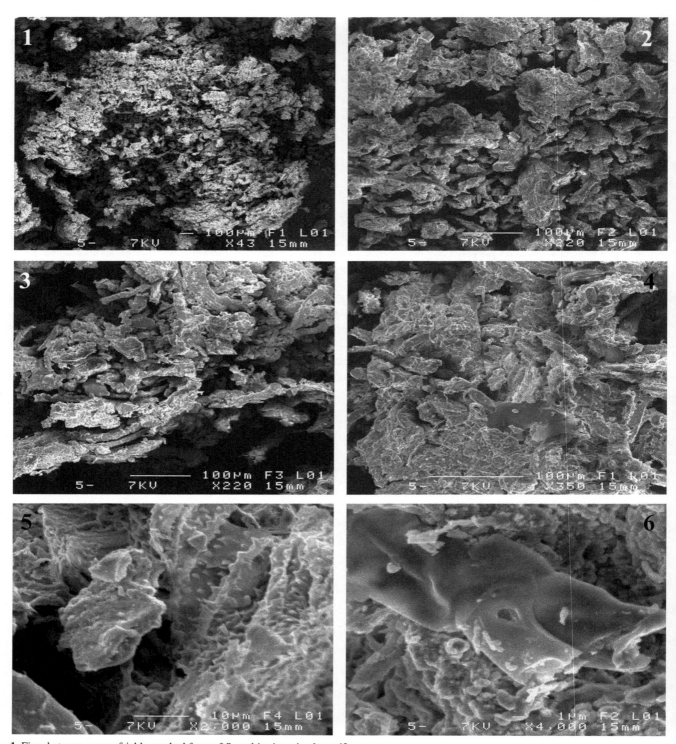

1. Fine, heterogeneous, friable, packed frass of *S. multistriatus* in elm, ×43
2. Frass with rare fusiform pellets, detail, ×220
3. Frass with wood fibres, detail, ×220
4. Poorly digested frass scraps, detail, ×350
5. Zoom on wood scraps. Semi-digested pieces of cell wall, ×2000
6. Zoom on wood fibres corresponding to undigested vessels and punctuation, ×4000

Scolytus rugulosus (Muller, 1818)

Fruit Bark Beetle

Description
The adult is stocky and measures 1.8–2.5 mm. It is dark brown, sometimes slightly reddish at the apex of its elytra. The punctuations of the elytra are very close to one another and sunken located at the bottom of small grooves.

Distribution
One generation is found at the latitude of northern France, and three in the hottest regions of the Mediterranean coast.

Bioecology
Adults can be seen from May to the end of July. The adult flies as soon as it comes out. When it has found a host plant, the female bores a single, vertical egg gallery in the same direction as the wood fibres, 20–30 mm long, at the edge of the sapwood and bark. Mating occurs several times during this drilling, with the male standing at the entrance of the gallery to keep rivals away. Oviposition takes 20–30 days, with 2–3 eggs *per* day. Average fecundity is 55 eggs.

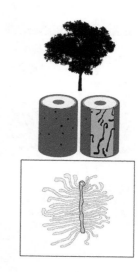

As soon as it hatches, the larva bores a gallery at the edge of the sapwood and bark on which it feeds. The larva is white with a brown head, apodic, curved and stocky. The larval galleries radiate out from the maternal gallery, are sinuous and often intersect at their ends. Their size increases as the larvae grow. Larvae develop during the spring and summer and, depending on the region, may pupate and start a new generation. As winter approaches, the larvae enter diapause in the pupation chamber. The larvae pupate in this small cavity located at the end of the gallery. Larvae that have not been able to make their way to this cavity die before pupation in March.

Galleries are mainly found in branches 3–4 cm in diameter located at the top of young trunks. The boring of the galleries impedes the flow of the sap and leads to the death of the host plant or organs. Beetle infestations are more frequent on trees already weakened by drought, cryptogamic diseases or infestations by animal pests.

Cycle of Development
The cycle of this species generally lasts 1 year.

Infested Woods
This bark beetle preferably grows on stone fruit trees of the genus *Prunus* such as plum, apricot, peach, cherry, more rarely on apple, pear, quince. It can also be found in the following species: elm (*Ulmus*), hawthorn (*Crataegus*), mountain ash (*Sorbus*), hazelnut (*Corylus*), birch (*Betula*) …

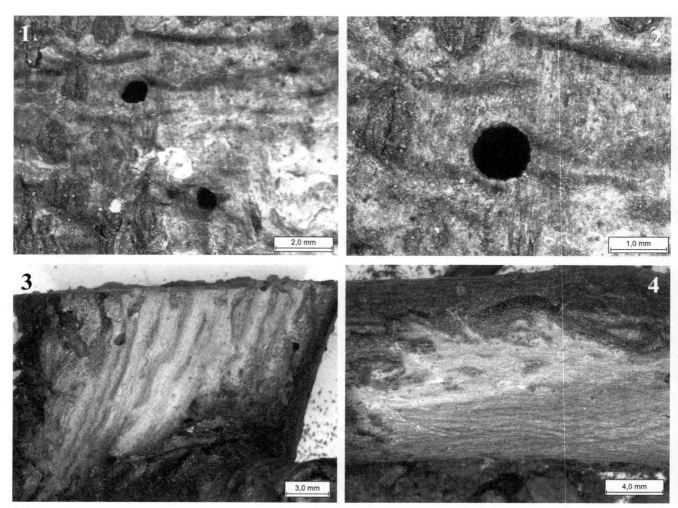

1–2. Sometimes very circular exit holes of *S. rugulosus* on a cherry tree (*Cerasus*) sample
3–4. Larval galleries located in the cambium and just penetrating the sapwood, detail

Maternal and larval galleries are irregularly shaped and circular to well rounded. Their diameter varies between 1 and 2 mm. The maternal gallery is simple, frass-less and in the same direction as the wood fibres (longitudinal). It spreads out under the bark, in the cambium of the tree along the sapwood but without actually penetrating it, over a length of 3–5 cm, sometimes more.

The larval galleries start at regular intervals from the egg-laying gallery. Then, each gallery (one gallery = one individual) is bored along a few centimetres (generally 3) through the cambium to end in a pupation chamber, which is slightly wider than the rest of the gallery. The adult will burrow through the bark from its chamber to emerge.

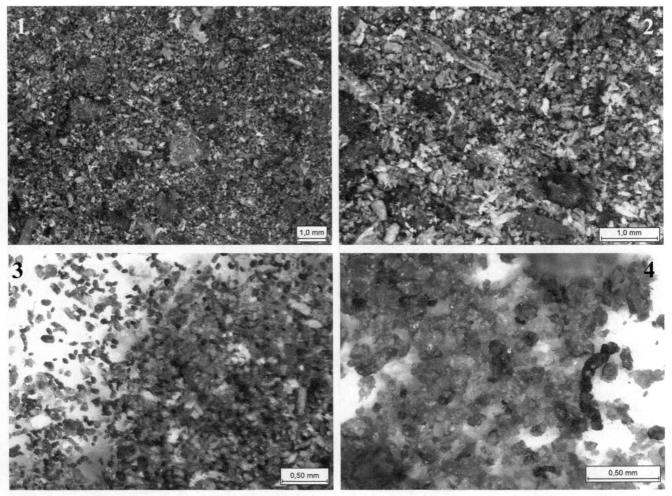

1–2. Granular, heterogeneous frass of *S. rugulosus,* general view
3–4. Heterogeneous frass with small, fusiform or rounded pellets mostly in relation to poorly digested wood scraps, detail

The frass is heterogeneous. It is composed of two assemblages: mostly 60–70% of small, fusiform to subspherical faecal pellets with rounded ends and rarely with a pointed apex. The pellet is generally no longer than 0.3 mm; and 30–40% of fine, undigested or poorly digested wood scraps (fibres). SEM observations show fungal hyphae along with the frass.

The frass is similar to that of other Scolytinae. Therefore, one should mind possible confusion with the frass of Cossoninae/Molytinae, which is also heterogeneous. Differentiation is made based on the typical galleries of the Scolytinae and their localisation in the wood.

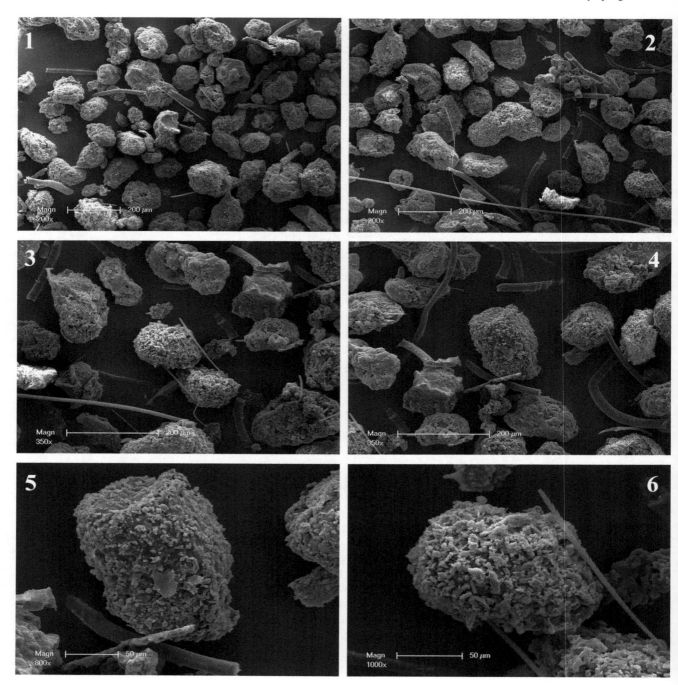

1. Heterogeneous frass of *S. rugulosus* on cherry tree (*Cerasus*), ×200
2. Shape variability of faecal pellet, ×200
3. A few pellets and partially digested wood debris, ×350
4. Fusiform pellet with fungal hyphae, ×350
5. Zoom on a pellet, fungal traces on the faecal pellet, ×800
6. Zoom on a rounded pellet and the wood fragment and fungal aggregates that adhered to it, ×1000

3.1.5 *Lyctidae*

Presentation of the Family

The family Lyctidae belongs to the super-family Bostrichoidea together with the families Bostrichidae and Ptinidae, to which Lyctidae are very close. Only one sub-family (Lyctinae) is found in Europe, with four genera (*Lyctoxylon*, *Lyctus*, *Minthea* and *Trogoxylon* for), and a total of 13 species altogether. The *Lyctus* genus is the largest one with nine species represented.

These insects are anatomically recognisable from their narrow, elongated bodies, usually between 2.5 and 7 mm long depending on the species, and more flattened than those of the other families. Their size depends on the quality of the feeding host.

They are strictly xylophagous beetles that prefer the sapwood of dry and/or worked wood. Thus, they are specialist insects of relatively dry dead wood and dried plant matter (roots of dry grasses).

Unlike the insects from the other two families mentioned above, their head is not embedded under their thorax so that the three constituent parts of the insect can be clearly distinguished. Moreover, their identification is facilitated by an antennal club consisting of only two items whereas the insects from the other two families have three.

The larvae are of the melolonthoid (curved) type, curved and enlarged at the level of their thorax, and are recognisable from the stigmas of their eighth abdominal segment which are larger than the others. They can be distinguished from those of Bostrichidae in particular because their head is more deeply embedded in their thorax.

Among the 13 species, six are well represented in Western Europe. This atlas focuses on the genera *Lyctus* and *Trogoxylon* (Table 3.5).

Lyctus and *Trogoxylon* can tolerate dry conditions in carbohydrate-rich environments. In addition, males produce aggregation pheromones when they find starchy products. These characteristics have enabled them to move from the natural environment to stored food products and to infest wood products, carvings and basketry, etc. and some have become harmful pests. Before laying their eggs, the females "taste" the wood by biting it to check that it is sufficiently rich in starch.

The degree of infestation and the speed of development of Lyctidae depend on:

- sapwood with large vessels ($\geq$0.05 mm in diameter), necessary for the laying of eggs and the first larval developments. Lyctidae only infest deciduous trees, preferably those with porous areas. Tropical woods are also likely to be infested by this family.
- the amount of starch present in the wood. The starch content of the wood must be more than 3% (w:w) of the weight of anhydrous wood. The larva targets starch-rich wood cells such as parenchymal areas, the pores of the initial wood at the beginning of its development, and then the transition from initial wood to final wood when the larva is bigger.
- the level of soluble sugars or so-called water-soluble substances: enzymes (some of which are still unknown) which are essential for the development of the larva, but which can be easily removed from the wood by immersing it in water at 60 °C.
- a high content in proteins and vitamins, known as nitrogenous substances. The tree is rich in them when the wood is cut. These levels decrease over time and make the feeding support less attractive. This is why worked wood is sensitive to Lyctidae during the first years of use of the wood.
- wood humidity between 7% and 32%, which is typical of timber. At a lower humidity rate, the larvae do not seem to be able to develop. Optimum ambient humidity varies between 40% and 60%.

Table 3.5

Species	Root	Branch	Under bark / Barkless	Trunk	Alive, declining/ dying	Decomposing	Newly fallen/ felled	Stored	Timber Deciduous	Coniferous	Ligneous species infested (non exhaustive list)	European distribution	Cycle	Gallerie diameter
Lyctus brunneus **(Stephens, 1830)**	x		x	x			x	x	x		Deciduous softwood with big vessels: *Castanea, Ficus, Fraxinus, Quercus, Ulmus* and tropical species	Austria, Azores Is., Belgium, Canary Is., Cyprus, Czech Republic, Denmark, Finland, France, Germany, Greece, Ireland, Italy, Malta, Norway, Poland, Portugal, Sardinia, Slovakia, Spain, Sweden, Switzerland, the UK	8–12 month Max 2 years	1–2 mm
Lyctus linearis **(Goeze, 1777)**	x		x	x			x	x	x		Deciduous softwood with big vessels: *Castanea, Ficus, Fraxinus, Quercus, Ulmus* and tropical species	Austria, Belgium, Bosnia and Herzegovina, Canary Is., Croatia, Czech Republic, Denmark, Finland, France, Germany, Hungary, Ireland, Italy, Latvia, Lithuania, Norway, Poland, Romania, Sardinia, Sicily, Slovakia, Spain, Sweden, Switzerland, The Netherlands, the UK	8–12 month Max 2 years	1–2 mm
Lyctus planicollis (Le Conte, 1858)	?		x	x			x	?	x		Deciduous woodland Little-known	Austria, Belgium, Germany, France, Madeira Is., Italy, The Netherlands, the UK	1 year	1–2 mm
Lyctus pubescens (Panzer, 1793)	?		x	x			x	?	x		Little-known	Austria, Belgium, Czech Republic, France, Germany, Hungary, Italy, Liechtenstein, Poland, Portugal, Romania, Sardinia, Slovakia, Spain, Switzerland, The Netherlands, Ukraine	1 year	1–2 mm

Species	Root	Branch	Under bark		Alive, declining/ dying	Decomposing	Newly fallen/ felled	Stored	Timber		Ligneous species infested (non exhaustive list)	European distribution	Cycle	Gallerie diameter
			Barkless	Trunk					Deciduous	Coniferous				
Minthea rugicollis (Walker, 1858)	x					?	?	?	x		Root systems	Belgium, Finland, France, Germany, Sweden, The Netherlands, the UK	1 year	1–2 mm
Trogoxylon impressum (Comolli, 1837)		x	x	x			x	x	x		Deciduous softwood with big vessels: *Castanea, Ficus, Fraxinus, Quercus, Ulmus* and tropical species	Austria, Balearic Is., Belgium, Canary Is., Croatia, Cyprus, Finland, France, Germany, Greece, Hungary, Italy, Norway, Romania, Sardinia, Sicily, Slovakia, Spain, Sweden, The Netherlands, Ukraine	Max 2 years	1–2 mm

Lyctus brunneus (Stephens, 1830)

Powder Post Beetle

Lyctus linearis (Goeze, 1777)

European Lyctus Beetle

Trogoxylon impressum (Comolli, 1837)

Powder Post Beetle

L. brunneus and *L. linearis* are relatively close anatomically and can be mistaken. In some aspects, the same applies to *Trogoxylon impressum*. Moreover, their respective biologies also remain similar.

Description

These reddish brown to dark brown insects are elongated and flattened in shape. They are 2.5–7 mm in length, depending on the nutritional value of the wood.

Anatomical differences are the following:

* the antennae are the same length as the pronotum in *L. brunneus* and longer than the pronotum in *L. linearis*.
* the prothorax is significantly wider at the front end than at the rear end in *L. brunneus*, whereas its edges are parallel in *L. linearis*.
* the elytra hairs are irregular in *L. brunneus*, whereas they are distributed in a longitudinal line in *L. linearis*.

Distribution

L. brunneus is cosmopolitan and present throughout the world. This exogenous species was first seen in Europe in the middle of the twentieth century; where it originally came from remains unclear. According to the literature, it could come from North America, Brazil or Asia. It is widely present in Western Europe.

L. linearis is an indigenous European species widely distributed in the western palearctic region, and is even found in some countries well-known to have rather cold climates such as Finland and Sweden.

T. impressum is a common species in the Mediterranean region. Its range extends to Central Europe (Switzerland, Germany, Austria) and it was also introduced into northern European countries and the USA, where it has become established.

Bioecology

Adult flight begins in spring and continues through summer.

After mating, females lay their eggs (about 20 eggs, singly or in groups of 2–6) in cracks or most often in the large wood vessels using their auger. Incubation lasts 1–2 weeks. The larvae then penetrate into the thickness of the material, where they bore numerous galleries. The length of larval development varies according to the thermo-hygrometric conditions. In winter, if conditions are bad, they remain inactive and thus prolong their development.

Infestation can only occur in woods that contain a high level of starch (optimum 3–4%) as well as other substances essential to their development such as soluble sugars, or certain vitamins and proteins. This characteristic makes them amylophagous or amylophilous insects.

The larva does not digest cellulose and hemicellulose A but digests starch, sucrose, maltose, lactose and hemicellulose B.

At the end of larval development, *Lyctus* transforms into a pupa in a small cavity that it builds for a period of 2–4 weeks. Between April and September, the adult finally emerges from the wood by boring a flight hole. It then adopts a twilight activity. The male usually lives for 2–8 weeks, compared to 6–16 weeks for the female (*L. linearis* has a better longevity).

Cycle of Development

These three species of Lyctidae have essentially the same life cycle.

The complete development takes about 8–10 months but can last up to 2 years in case of unfavourable thermo-hygrometric conditions or, conversely, can go down to less than 6 months under optimal breeding conditions in the laboratory.

The duration of the complete life cycle of these Lyctidae is therefore usually 1 year, with rarely more than one generation.

Infested Woods

Lyctidae infest dry and/or worked wood mainly in synanthropic environments (workshops, plywood factories …). In the wild, the species has been found in alder (*Alnus*), eucalyptus (*Eucalyptus*), elm (*Ulmus*), all presenting poor sanitary conditions.

The softwood and sapwood parts of deciduous species such as oak (*Quercus*), walnut (*Juglans*), ash (*Fraxinus*), elm (*Ulmus*), cherry (*Cerasus*), chestnut (*Castanea*), black locust (*Robinia*) … are infested. These infestations are rarer in woods with large vessels such as apricot (*Prunus armeniaca*), cherry (*Cerasus*), alder (*Alnus*), birch (*Betula*), olive (*Olea*) (rich in parenchyma), poplar (*Populus*), apple (*Malus*), willow (*Salix*) … Tropical species rich in starch such as mahogany (*Swietenia* spp.), bamboo (Bambusoideae), acacia (*Acacia*), koto (*Pterygota bequaertii*), ayous (*Triplochiton scleroxylon*) are also widely infested. However, conifers are not infested. Finally, other species such as beech (*Fagus*), hornbeam (*Carpinus*), almond (*Prunus dulcis*), quince (*Cydonia oblonga*), mulberry (*Morus*), plum (*Prunus*) and linden (*Tilia*) seem to be resistant.

Trogoxylon impressum has been reported on various woods in addition to dry worked woods including Carob tree (*Ceratonia siliqua*), Judah tree (*Cercis siliquastrum*), eucalyptus (*Eucalyptus* spp.), fig (*Ficus carica*), pistachio tree (*Pistacia lentiscus, P. vera*), pomegranate (*Punica granatum*), oak (*Quercus* spp.), tamarisk (*Tamarix*) and vine tree (*Vitis vinifera*).

1. Oak (*Quercus*) sample infested by the genus *Lyctus*
2. Same oak sample, experimentally charred
3. *Lyctus* galleries, location and targeting of initial wood by the larvae, detail
4. *Lyctus* galleries, location and targeting of visible cells, on thin coloured slide, cl. A. Crivellaro
5. Galleries of *Lyctus brunneus* in tangential view
6. Galleries of *Trogoxylon impressum* in tangential view

The larval galleries of Lyctidae are very round, and 0.8–1.5 to 2 mm in diameter.

They are filled with frass, so that infestation is often not detected until very late. The wood can be completely destroyed over several generations because re-infestation occurs many times on the same wood.

The larvae bore galleries, usually in the same direction as the grain of the wood, and target certain starch-rich cells such as the initial wood of oak (with large vessels in spring).

At first glance, the galleries of Lyctidae can be mistaken for those of Ptinidae. The main characteristics to differentiate them is the targeting of wood cells which Ptinidae do not do. Next, it is necessary to study the frass.

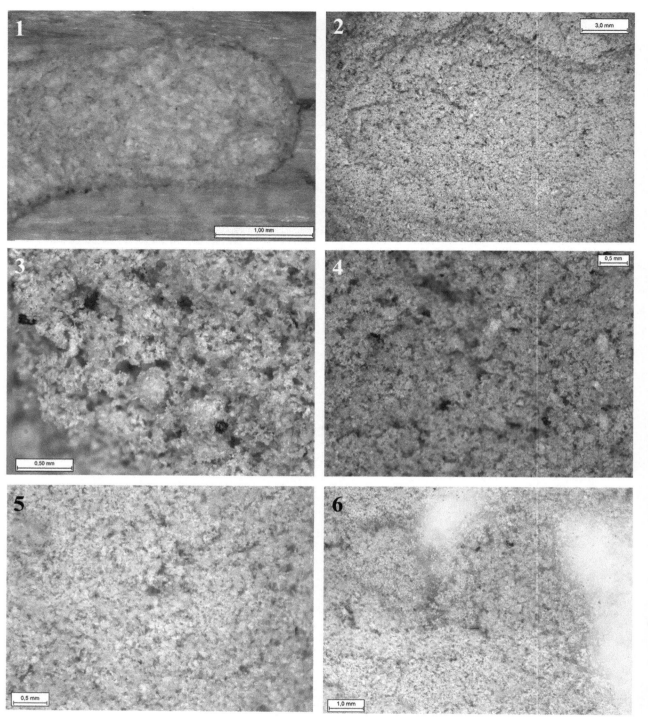

1. Frass visible in a gallery
2–3. Fine lyctid frass
4. Lyctid frass, detail
5–6. Frass of *Trogoxylon impressum*

The frass is homogeneous and composed of a multitude of fibres and cell walls poorly digested or undigested by the larva. It is identical for the three species.

It is a very fine, powder-like frass, regularly called fine-flour type in the literature. No particular pellet is discernible.

The shape and size of the galleries of Ptinidae can be mistaken for those of Lyctidae. Ptinidae produce a granular and fusiform frass resembling faecal pellets. Bostrichidae, which have similar frass, bore galleries of a greater diameter.

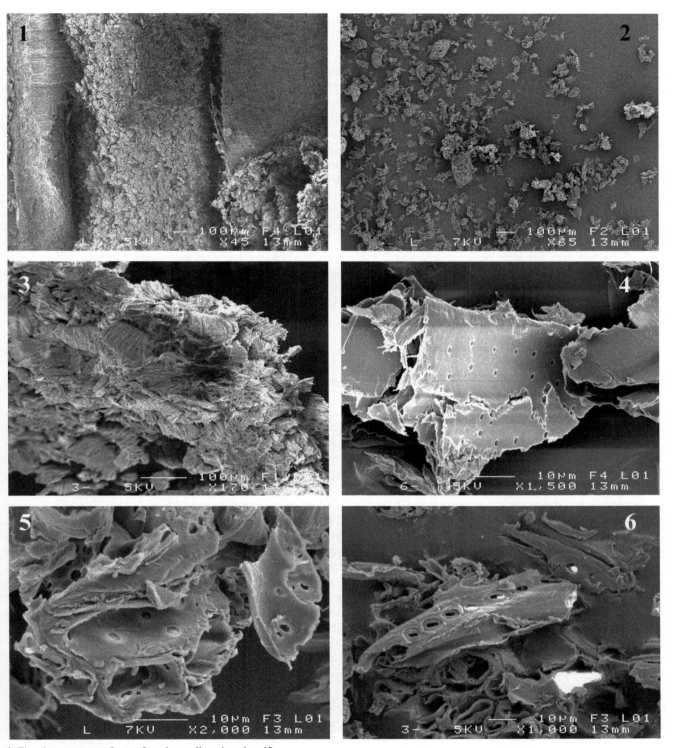

1. Fine, homogeneous *Lyctus* frass in a gallery, in oak, ×45
2. Frass of *Lyctus*, oak sample, ×65
3. Wood fibres, more or less agglomerated but not forming distinct clumps, ×170
4. Wood fibers, undigested cells, *Trogoxylon*, ×1500
5. Wood fibers, undigested cells, ×2000
6. Wood fiber, undigested cells, *Lyctus*, ×1000

3.1.6 Ptinidae

Presentation of the Family

This family was formerly at the subfamily level (Ptininae), but now includes all the subfamilies previously included in the family Anobiidae. Despite this taxonomic switch, a number of species of the former family Anobiidae will be considered in this book. Ptinidae, which belong to the super-family Bostrichoidea in the same way as the family Bostrichidae to which they are very close, gather eight subfamilies designated under the term "beetle" (Anobiinae, Dorcatominae, Dryophilinae, Ernobiinae, Eucradinae, Mesocoleopodinae, Ptilininae, Xyletininae) and two subfamilies (Gibbiinae, Ptininae), many of whose representatives are synanthropic and live in human dwellings where they can cause a lot of damage to stored food.

Ptinidae are found throughout the world and generally appear to be more diverse in temperate zones than in tropical areas. There are about 230 genera and more than 2200 known species worldwide. Their habits are extremely diverse for a group of such a small size.

This family of beetles does not only include xylophagous species but also pests of stored goods, predators or detritivores (*Stegobium paniceum, Lasioderma serricorne, Xyletinus* spp.) (not dealt with in the present book), fungivores (*Anitys* spp., *Caenocara* spp., *Dorcatoma* spp., *Stegatus* spp.) (not dealt with in the present book), cambiophagous species (*Dryophilus* spp., *Episernus* spp., *Ernobius mollis, E. abietinus*), xylemophagous species (*Anobium punctatum, Calymmaderus solidus, Nicobium castaneum, Oligomerus ptilinoides, Xestobium rufovillosum*) and maybe even more saproxylophagous species (*Grynobius planus, Hadrobregmus pertinax, Hemicoelus* spp., *Microbregma emarginatum, Ptilinus pectinicornis, Ptinomorphus imperialis, Ochina latrellii, Xestobium declive*). The subfamilies Gibiinae and Ptininae are not included in this book because very few species are woodborers and a number of them are found in stored food.

Some of the woodborers, or "woodworms", are among the most well-known pests in the world because of their high economic impact. In Western Europe, there are currently about 60 harmful and often cosmopolitan xylophagous Ptinidae; the most famous of them are the common furniture beetle (*Anobium punctatum*), the deathwatch beetle (*Xestobium rufovillosum*) and the brown woodworm (*Oligomerus ptilinoides*). All three are great pests of woodwork, furniture and heritage objects.

These beetles are small (2–8 mm) and can be recognised thanks to the presence of a "cap" covering their head (prothorax), which gives them their typical overall shape. Adults are more or less cylindrical, with antennae composed of 11 items, the last three of which are more elongated and form the antennal club. Many of these species host fungi in which symbionts (mainly bacteria but also yeasts) are hosted, providing them with essential nutrients for their development.

The larvae of Ptinidae are small (about 10 mm), curved, pale and hairless or with scattered bristles. Their head is small, darker than the body. Their mandibles are well developed, symmetrical, triangular and pointed.

In general, woodworm larvae develop in dead, freshly felled wood, and more particularly in dry wood, which they can totally deteriorate. Host trees/woods can be deciduous trees, coniferous trees or both, depending on the species: *Ernobius* spp. on *Pinus*; *Dryophilus pusillus* on conifers; *Ptilinus pectinicornis* on *Fagus, Populus, Quercus* and *Salix*; *Hadrobregmus denticollis, Dryophilus longicollis* on deciduous trees; *Ochina ptinoides* on *Hedera*; *Hadrobregmus pertinax* on both deciduous and coniferous trees; *Anobium* ssp, *Oligomerus* ssp, *Hemicoelus* ssp, on various species. Adults do not feed and live only a few weeks. Oviposition usually takes place in cracks in the wood or in old emergence holes. The larvae then bore into the wood and sometimes reach the heartwood when the wood has been previously degraded by lignivorous fungi. The length of the evolutionary cycle varies according to thermo-hygrometric factors and the nutritional value of the wood. If conditions are not favourable enough, the cycle lasts longer, and the larvae then fall into an inactive state (Table 3.6).

Table 3.6

Species	Root	Branch	Under bark	Barkless	Trunk	Alive, declining/ dying	Decomposing	Newly fallen/ felled	Stored	Timber	Deciduous	Coniferous	Ligneous species infested (non exhaustive list)	European distribution	Cycle	Gallerie diameter
Anobium hederae (Ihssen, 1949)		x			x			x	x		x		*Hedera*	Austria, Belgium, Bulgaria, Croatia, Czech Republic, France, Germany, Liechtenstein, Portugal, Slovenia, Spain, Switzerland	>1 year ?	1–3 mm
Anobium inexpectatum (Lohse, 1954)		x			x	x		x	x		x		*Hedera, Quercus*	Austria, Belgium, France, Germany, Slovenia, Spain, Switzerland, The Netherlands, the UK	>1 year ?	1–3 mm
Anobium punctatum (De Geer, 1774)		x		x	x			?	x	x	x	x	**Various**	**More common in Western Europe**	**1–4 years Until 10 years**	**1–3 mm**
Cacotemnus rufipes (Fabricius, 1792)		x		?	x			?	x		x	x	Little-known	Austria, Belgium, Croatia, Czech Republic, Finland, France, Germany, Hungary, Italy, Latvia, Lithuania, Norway, Poland, Romania, Slovakia, Sweden, Switzerland, The Netherlands, the UK, Ukraine	>1 year ?	1–3 mm
Calymmaderus solidus (Kiesenwetter, 1877)		x		x	x			?	x	x	x	?	**Little-known**	**France, Italy, Portugal, Spain**	**>1 year ?**	**1–3 mm**
Dryophilus anobioides (Chevrolat, 1832)	x	x	x		x	x		x	x		x		*Genista, Ulex*	Austria, Belgium, Croatia, Czech Republic, France, Germany, Hungary, Italy, Moldova, Poland, Sardinia, Sicily, Slovakia, Slovenia, Spain, Switzerland, The Netherlands, the UK, Ukraine	>1 year ?	1–3 mm
Dryophilus densipilis (Abeille, 1872)	x	x	x		x	x	?	x	x			x	Little-known	Mediterranean basin: France, Greece, Italy, Portugal, Sardinia, Sicily, Spain	>1 year ?	1–3 mm

(continued)

Table 3.6 (continued)

Species	Root	Branch	Under bark	Barkless	Trunk	Alive, declining/ dying	Decomposing	Newly fallen/ felled	Stored Timber	Deciduous	Coniferous	Ligneous species infested (non exhaustive list)	European distribution	Cycle	Gallerie diameter
Dryophilus longicollis (Mulsant and Rey, 1853)		x	x		x	x	?	x	x	x		Little-known	Austria, Croatia, Czech Republic, France, Germany, Greece, Italy, Poland, Sicily, Slovakia, Spain, Switzerland, Ukraine	>1 year ?	1–3 mm
Dryophilus pusillus (Gyllenhal, 1808)		x	x		x	x	x	x	x		x	*Abies, Larix, Picea, Pinus*	Common in Western Europe	>1 year ?	1–3 mm
Episernus gentilis (Rosenhauer, 1847)		x	x		x			x	x	x		*Laburnum*	Austria, France, Germany, Italy, Sardinia, Switzerland	>1 year ?	1–3 mm
Episernus hispanus (Kiesenwetter, 1877)		x	x		x			x	x	x	?	Little-known	France, Spain	>1 year ?	1–3 mm
Ernobius abietinus (Gyllenhal, 1808)		x	x		x	?		x	x		x	*Abies, Pinus*	Austria, Belgium, Croatia, Czech Republic, Denmark, Estonia, Finland, France, Germany, Greece, Italy, Latvia, Norway, Poland, Romania, Slovakia, Sweden, Switzerland, The Netherlands, Ukraine	>1 year ?	1–3 mm
Ernobius abietis (Fabricius, 1792)		x	x		x	?		x	x		x	*Abies, Pinus*	Austria, Belgium, Bulgaria, Croatia, Czech Republic, Denmark, Estonia, Finland, France, Germany, Greece, Italy, Latvia, Liechtenstein, Lithuania, Norway, Poland, Romania, Slovakia, Slovenia, Sweden, Switzerland, The Netherlands, the UK, Ukraine	>1 year ?	1–3 mm

Species									Host plant	Distribution	Lifespan	Size
Ernobius angusticollis (Ratzeburg, 1837)	x	x	?	x	x			x	x	Mainly: *Pinus*	Austria, Belgium, Czech Republic, Denmark, Finland, France, Germany, Greece, Hungary, Italy, Latvia, Lithuania, Norway, Poland, Romania, Slovakia, Slovenia, Sweden, Switzerland, The Netherlands, the UK	>1 year ?	1–3 mm
Ernobius cupressi (Chobaut, 1899)	x	x	?	x	x			x	x	Mainly: *Cupressaceae*	Cyprus, France, Greece, Spain	>1 year ?	1–3 mm
Ernobius explanatus (Mannerheim, 1843)	x	x	?	x	x			x	x	Mainly: *Pinus*	Central and Northern Europe	>1 year ?	1–3 mm
Ernobius gallicus (Johnson, 1975)	x	x	?	x	x			x	x	Mainly: *Pinus*	France, Spain	>1 year ?	1–3 mm
Ernobius gigas (Mulsant and Rey, 1863)	x	x	?	x	x			x	x	Mainly: *Pinus*	Balearic Is., France, Germany, Latvia, Lithuania, Portugal, Spain, Switzerland, the UK, Ukraine	>1 year ?	1–3 mm
Ernobius juniperi (Chobaut, 1899)	x	x	?	x	x			x	x	Mainly: *Pinus*	France, Sardinia, Spain	>1 year ?	1–3 mm
Ernobius kiesenwetteri (Schilsky, 1898)	x	x	?	x	x			x	x	Mainly: *Pinus*	Austria, Croatia, Czech Republic, France, Greece, Italy, Latvia, Lithuania, Macedonia, Poland, Romania, Sardinia, Switzerland	>1 year ?	1–3 mm
Ernobius laticollis (Pic, 1927)	x	x	?	x	x			x	x	Mainly: *Pinus*	Austria, France, Greece, Italy, Sicily, Spain	>1 year ?	1–3 mm
Ernobius longicornis (Sturm, 1837)	x	x	?	x	x			x	x	Mainly: *Pinus*	Austria, Belgium, Bosnia and Herzegovina, Czech Republic, Denmark, Estonia, Finland, France, Germany, Italy, Latvia, Lithuania, Norway, Poland, Romania, Slovakia, Sweden, Switzerland, The Netherlands	>1 year ?	1–3 mm

(continued)

152 3 Atlas of the Most Common Xylophagous Insects

Table 3.6 (continued)

Species	Root	Branch	Under bark	Barkless	Trunk	Alive, declining/dying	Decomposing	Newly fallen/felled	Stored	Timber	Deciduous	Coniferous	Ligneous species infested (non exhaustive list)	European distribution	Cycle	Gallerie diameter
Ernobius lucidus (Mulsant and Rey, 1863)		x	x		x	?		x	x			x	Mainly: Pinus	Balearic Is., Bulgaria, France, Germany, Italy, Portugal, Spain	>1 year ?	1–3 mm
Ernobius mollis (Linnaeus, 1758)		**x**	**x**		**x**	**x**		**x**	**x**	**x**		**x**	**Mainly: Pinus, Cupressus**	**More common in Western Europe**	**>1 year ?**	**1–3 mm**
Ernobius mulsanti (Kiesenwetter, 1877)		x	x		x	?		x	x			x	Mainly: Pinus	Austria, France, Germany, Italy, Poland	>1 year ?	1–3 mm
Ernobius nigrinus (Sturm, 1837)		x	x		x	?		x	x			x	Mainly: Pinus	Common in Western Europe	>1 year ?	1–3 mm
Ernobius parens (Mulsant and Rey, 1863)		x	x		x	?		x	x			x	Mainly: Pinus	Balearic Is., France, Germany, Hungary, Italy, Portugal, Sardinia, Sicily, Spain, Ukraine	>1 year ?	1–3 mm
Ernobius pini (Sturm, 1837)		x	x		x	?		x	x			x	Mainly: Pinus	Austria, Belgium, Croatia, Cyprus, Czech Republic, Denmark, Estonia, Finland, France, Germany, Greece, Hungary, Italy, Latvia, Poland, Romania, Sardinia, Sicily, Slovakia, Spain, Sweden, Switzerland, The Netherlands, the UK	>1 year ?	1–3 mm
Ernobius pruinosus (Mulsant and Rey, 1863)		x	x		x	?		x	x			x	Mainly: Pinus	Balearic Is., France, Germany, Greece, Italy, Slovakia, Spain	>1 year ?	1–3 mm
Ernobius reflexus (Mulsant and Rey, 1863)		x	x		x	?		x	x			x	Mainly: Pinus	Balearic Is., France, Germany, Italy, Spain	>1 year ?	1–3 mm
Ernobius rufus (Illiger, 1807)		x	x		x	?		x	x			x	Mainly: Pinus	France, Portugal, Sardinia, Spain	>1 year ?	1–3 mm
Gastrallus corsicus (Schilsky, 1898)			x		x	?	?	x	x		x		Little-known	Mediterranean basin: Balearic Is., Croatia, France, Greece, Italy, Portugal, Sicily, Spain	>1 year ?	1–3 mm

Species									Host	Distribution	Life cycle	Size
Gastrallus immarginatus (P. W. J. Muller, 1821)	x	x		x	?	x	x	?	*Acer*	Austria, Belgium, Croatia, Czech Republic, Denmark, France, Germany, Greece, Hungary, Italy, Poland, Romania, Sardinia, Sicily, Slovakia, Spain, Sweden, Switzerland, the UK, Ukraine	>1 year ?	1–3 mm
Gastrallus knizeki (Zahradnik, 1996)	x	x		x	?	x	x		*Rosaceae, Tillia, Viscum*	Austria, Czech Republic, France, Germany, Slovakia	>1 year ?	1–3 mm
Gastrallus laevigatus (Olivier, 1790)	x		x	x	?	x	x		Various	Common in Western Europe	>1 year ?	1–3 mm
Grynobius planus (Fabricius, 1787)	x	x	x	x	x	x	x		Various	Austria, Balearic Is., Belgium, Denmark, Finland, France, Germany, Hungary, Ireland, Italy, Poland, Spain, Sweden, Switzerland, The Netherlands, the UK, Ukraine	>1 year ?	1–3 mm
Hadrobregmus bicolor (Español, 1990)	x		x	x	x	x	x		Various	France, Spain	>1 year ?	1–3 mm
Hadrobregmus denticollis (Creutzer, 1797)	x		x	x	?	x	?		Various	Austria, Belgium, Czech Republic, Denmark, France, Germany, Hungary, Italy, Latvia, Lithuania, Poland, Romania, Slovakia, Spain, Sweden, Switzerland, The Netherlands, the UK	>1 year ?	1–3 mm
Hadrobregmus pertinax (Linnaeus, 1758)	x		x	x	x	x	?	x	Various	Common in Western Europe	>1 year ?	1–3 mm
Hedobia pubescens (Olivier, 1790)	x		x	x	?	x	?		Various	Austria, Croatia, Czech Republic, France, Greece, Hungary, Italy, Poland, Romania, Sardinia, Spain, Slovakia, Ukraine	>1 year ?	1–3 mm

(continued)

Table 3.6 (continued)

Species	Root	Branch	Under bark	Barkless	Trunk	Alive, declining/ dying	Decomposing	Newly fallen/ felled	Stored	Timber	Deciduous	Coniferous	Ligneous species infested (non exhaustive list)	European distribution	Cycle	Gallerie diameter
Hemicoelus costatus (Aragona, 1830)		x			x		x	x	x		x	x	Various	Albania, Austria, Belgium, Bulgaria, Croatia, Czech Republic, Denmark, France, Germany, Hungary, Italy, Liechtenstein, Macedonia, Poland, Portugal, Romania, Sardinia, Sicily, Slovakia, Slovenia, Spain, Sweden, Switzerland, The Netherlands, Ukraine	>1 year ?	1–3 mm
Hemicoelus fulvicornis (Sturm, 1837)		x			x		x	x	x		x	x	Various	Austria, Balearic Is., Belgium, Bulgaria, Croatia, Czech Republic, Denmark, Finland, France, Germany, Greece, Hungary, Italy, Poland, Romania, Sardinia, Slovakia, Slovenia, Spain, Sweden, Switzerland, The Netherlands, the UK, Ukraine	>1 year ?	1–3 mm
Hemicoelus nitidus (Fabricius, 1792)		x			x		x	x	x		x	x	Various	Austria, Balearic Is., Belgium, Bulgaria, Croatia, Czech Republic, Denmark, Estonia, Finland, France, Germany, Greece, Italy, Latvia, Malta, Norway, Poland, Romania, Sardinia, Slovakia, Spain, Sweden, Switzerland, The Netherlands, the UK, Ukraine	>1 year ?	1–3 mm

Species								Diet	Distribution	Development	Size
Hemicoelus rufipennis (Duftschmid, 1825)	x		x	x	x	x	x	Little-known	Austria, Bulgaria, Croatia, Czech Republic, France, Germany, Greece, Italy, Poland, Romania, Slovakia	>1 year ?	1–3 mm
Mesocoelopus collaris (Mulsant and Rey, 1864)	x		x	?	x	?	x	Little-known	Austria, Balearic Is., Croatia, France, Germany, Italy, Malta, Portugal, Sardinia, Sicily, Spain, the UK	>1 year ?	1–3 mm
Mesocoelopus niger (Müller, 1821)	x		x	?	x	?	x	Little-known	Austria, Belgium, Croatia, Czech Republic, France, Germany, Greece, Italy, Liechtenstein, Poland, Portugal, Sardinia, Sicily, Slovakia, Spain, Switzerland, Ukraine	>1 year ?	1–3 mm
Metholcus phoenicis (Fairmaire, 1859)	x		x	?	x	?	x	Little-known	Mediterranean basin	>1 year ?	1–3 mm
Microbregma emarginatum (Duftshmid, 1825)	x		x	x	x	x	x	Little-known	Austria, Belgium, Czech Republic, Finland, France, Germany, Hungary, Italy, Latvia, Lithuania, Norway, Poland, Portugal, Romania, Slovakia, Spain, Sweden, Switzerland, The Netherlands	>1 year ?	1–3 mm
Nicobium castaneum (Olivier, 1790)	x	x	x	x	x	x	?	Various	**Austria, Azores Is., Balearic Is., Bosnia and Herzegovina, Canary Is., Croatia, Cyprus, Czech Republic, France, Germany, Greece, Italy, Madeira Is., Malta, Portugal, Romania, Sardinia, Sicily, Slovenia, Spain, Switzerland, Ukraine**	**1–3 years Until 10 years**	**1–3 mm**
Ochina hirsuta (Seidlitz, 1889)	x		x	?	x	?	x	Little-known	France, Italy, Sardinia, Spain	>1 year ?	1–3 mm

(continued)

Table 3.6 (continued)

Species	Root	Branch	Under bark	Barkless	Trunk	Alive, declining/dying	Decomposing	Newly fallen/felled	Stored	Timber	Deciduous	Coniferous	Ligneous species infested (non exhaustive list)	European distribution	Cycle	Gallerie diameter
Ochina latreillii (Bonelli, 1809)		x			x		?	x	?		x	?	Little-known	Austria, Belgium, Croatia, Czech Republic, France, Germany, Greece, Hungary, Italy, Romania, Sardinia, Slovakia, Spain, Switzerland, Ukraine	>1 year ?	1–3 mm
Ochina ptinoides (Marsham, 1802)		x			x		?	x	?		x		*Hedera*	Austria, Belgium, Croatia, Czech Republic, Denmark, France, Germany, Greece, Hungary, Ireland, Italy, Latvia, Liechtenstein, Lithuania, Portugal, Sardinia, Sicily, Spain, Switzerland, The Netherlands, the UK	>1 year ?	1–3 mm
Oligomerus brunneus (Olivier, 1790)		x		x	x			x	x	x	x		Little-known	Austria, Balearic Is., Belgium, Bulgaria, Croatia, Czech Republic, France, Germany, Hungary, Italy, Latvia, Lithuania, Poland, Romania, Sardinia, Sicily, Slovakia, Slovenia, Spain, Sweden, Switzerland, Ukraine	>1 year ?	1–3 mm
Oligomerus ptilinoides (Wollaston, 1854)		**x**		**x**	**x**			**x**	**x**	**x**	**x**		**Various**	**Austria, Balearic Is., Canary Is., Croatia, Cyprus, France, Germany, Greece, Hungary, Italy, Madeira Is., Malta, Poland, Portugal, Romania, Sardinia, Sicily, Slovakia, Slovenia, Spain, Switzerland, Ukraine**	**1 year Until 3 years**	**1–3 mm**
Pseudodryo philus paradoxus (Rosenhauer, 1856)		x			x		?	x	?		x		Little-known	France, Greece, Italy, Spain	>1 year ?	1–3 mm

Species								Host	Distribution	Development	Size
Priobium carpini (Herbst, 1793)		x	?	x	x		x	Various	Common in Western Europe	>1 year ?	1–3 mm
Ptilinus fuscus (Geoffroy, 1785)		x	x	x	x		x	Little-known	Austria, Belgium, Bulgaria, Czech Republic, Estonia, Finland, France, Germany, Greece, Hungary, Italy, Macedonia, Norway, Poland, Sardinia, Sicily, Slovakia, Spain, Sweden, Switzerland, The Netherlands, the UK, Ukraine	>1 year ?	1–3 mm
Ptilinus pectinicornis (Linnaeus, 1758)		x	x	x	x		x	*Fagus, Populus, Quercus, Salix*	Common in Western Europe	>1 year ?	1–3 mm
Ptinomorphus angustatus (Brisout, 1862)		x	x	x	x		x	Little-known	France, Italy, Spain	>1 year ?	1–3 mm
Ptinomorphus imperialis (Linnaeus, 1767)		x	x	x	x		x	Little-known	Common in Western Europe	>1 year ?	1–3 mm
Ptinomorphus regali (Duftschmid, 1825)		x	x	x	x		x	Little-known	Common in Western Europe	>1 year ?	1–3 mm
Xestobium declive (Dufour, 1843)		x	?	x	x		x	Various	France, Italy, Spain, Switzerland	>1 year ?	>3 mm
Xestobium plumbeum (Illiger, 1801)		x	?	x	x		x	Various	Austria, Belgium, Bosnia and Herzegovina, Bulgaria, Croatia, Czech Republic, France, Germany, Greece, Hungary, Italy, Macedonia, Poland, Romania, Slovakia, Slovenia, Spain, Switzerland, The Netherlands, the UK, Ukraine	>1 year ?	>3 mm
Xestobium rufovillosum (De Geer, 1774)	x	x	x	x	x	x	x	**Various**	**Common in Western Europe**	**2–3 years Until 10 years**	**>3 mm**

Anobium punctatum (De Geer, 1774)

Common Furniture Beetle

Description
More or less dark brown depending on its feeding medium, *Anobium punctatum* has an elongated shape and is 2–5 mm long. This species can be distinguished from the other Ptinidae based on the shape of its cap consisting of a big triangular median protuberance, as well as its slightly granular, striated elytra punctuated by large deep dots arranged in regular lines.

The larva is arched and composed of ten segments. It is 4–5 mm long and up to 2 mm wide.

Distribution
Cosmopolitan or even anthropophilous, but with a preference for temperate climate zones. It is very frequently found in houses and museums.

Bioecology
This species can be found from spring until September. Adults live 4 weeks on average. It infests woods when the ambient temperature reaches 17 °C.

Larval development is optimal at a humidity rate of around 30% for wood or 55–65% for ambient humidity. This is possible between 12 and 29 °C, with an optimum at 22 °C.

Each female lays 10–40 eggs on average in small batches in wood cracks and old exit holes. Incubation and pupation last about 2 weeks each.

Prior infestation of the wood by lignivorous fungi favours the establishment of these insects, and damage is much greater because it is then easier for the larvae to digest and attack the harder areas of the wood.

The larva does not build a pupation shell but occupies a space below the surface of the wood.

Cycle of Development
The development cycle varies from 1 to 4 years depending on thermo-hygrometric conditions. It can be greatly shortened (a few months) in the case of a fungal attack.

Conversely, in extreme conditions, the larva can "hibernate" for up to 10 years. There can be two annual generations in warm climates.

Infested Woods
This species particularly infests dry and manufactured wood, from both deciduous and coniferous trees. It is the furniture woodworm par excellence. It will favour the sapwood but can also infest the heartwood if previously degraded by a lignivorous fungus.

Alder (*Alnus*), birch (*Betula*), walnut (*Juglans*), pine (*Pinus*), poplar (*Populus*), chestnut (*Castanea*) and fir (*Abies*) are particularly affected. Infestations of oak and ash and some exotic species seem rarer.

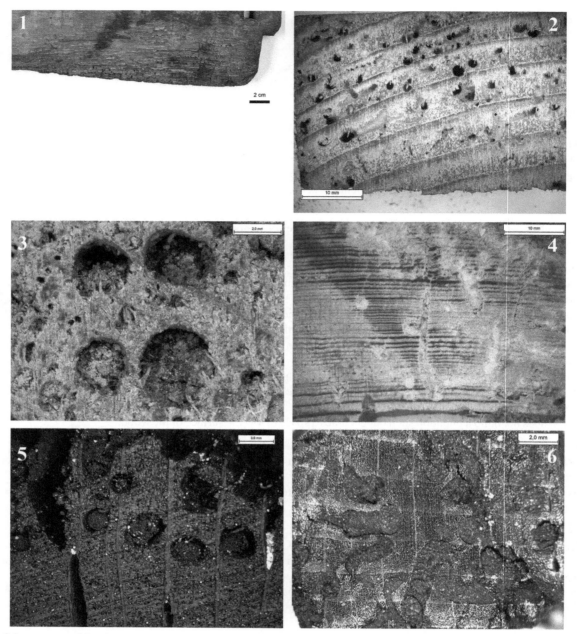

1. View of damage caused by *A. punctatum* on an archaeological object (Roman period, roof frame web, site of Rezé, France) made of oak (*Quercus*) (waterlogged wood)

2–3. Galleries in fir tree (*Abies*)

4. Galleries in Scots pine (*Pinus sylvestris*)

5. Galleries in an experimentally charred sample of oak (*Quercus*)

6. Galleries of *A. punctatum* on an archaeological beech charcoal, Camelin site (Fréjus, France)

The galleries of *Anobium punctatum* are round and quite regular, with a diameter ranging from 1.5 to 3 mm maximum (average: 1.9 mm). The larvae do not target specific cells. This makes a difference with Lyctidae, a family with which woodworms can be confused. When the generations follow one another on a same object or host, the larval galleries intertwine. The galleries are bored preferentially in the same direction as the grain of the wood, but easily accommodate to obstacles (knots, cracks, varnish, paint …).

Because of their size and circular shape, the adults' flight holes can be mistaken for those of other Ptinidae such as *Calymmaderus solidus*, *Oligomerus ptilinoides* or *Nicobium castaneum*. The distinction between these species is then made based on the macro- and micro-characteristics of the frass.

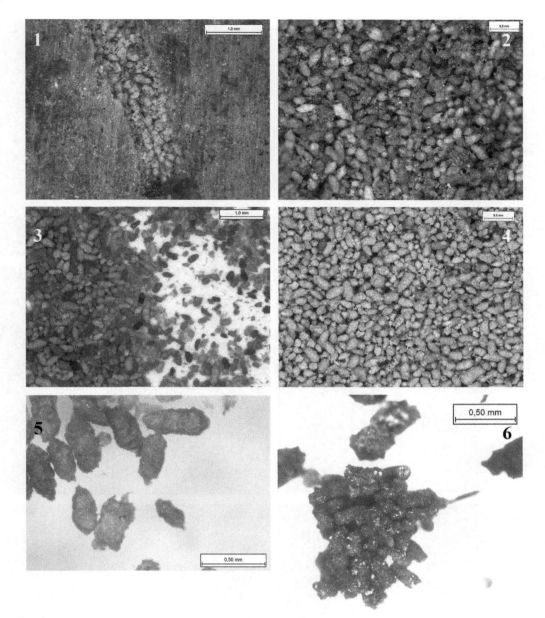

1. Frass of *A. punctatum* in a gallery, overview
2. Frass in pine (*Pinus sylvestris*), overview
3. Gallery and faecal pellet in fir tree (*Abies*)
4. Gallery and faecal pellet in oak (*Quercus*)
5. Fusiform pellets, detail
6. Frass from archaeological beech (*Fagus*) charcoal, archaeological site of Camelin (Fréjus, France)

The frass of *A. punctatum* is homogeneous and granular and consists of a multitude of fusiform faecal pellets that have a general peanut-like aspect.

The size depends on the larval stage; each pellet is about 0.33 mm long by 0.15 mm wide on average. There is no significant difference in the aspect of the frass depending on the infested woody species.

Each pellet is exclusively composed of more or less well digested wood. SEM photographs illustrate the anarchic arrangement of wood scraps within a fusiform pellet.

The details of the pellets are variable, and the ends of a fusiform pellet are not characteristic. One end (apex) may be sharp, or both.

The pellets are brittle, and may end up packed with the rest of the frass.

The frass of *A. punctatum* can be mistaken for the frass of *Calymmaderus solidus*, *Oligomerus ptilinoides* and *Nicobium castaneum*. It is necessary to refer to average measurements and the frequency of details to differentiate between them: in the Ptinidae group, *A. punctatum* produces the smallest and most irregular faecal pellet as compared to the other species.

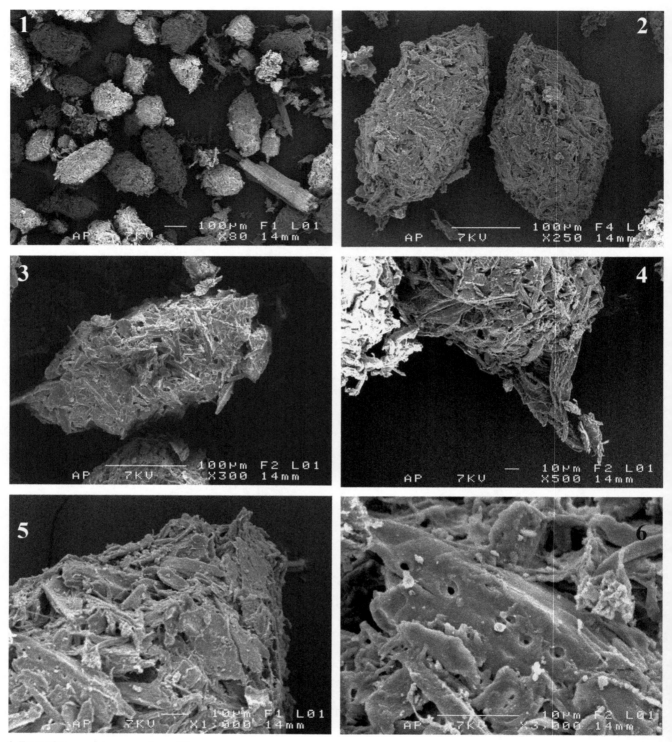

1. Frass of *A. punctatum*, overview, ×80
2. Two fusiform pellets, ×250
3. More or less digested wood scraps, detail and arrangement, ×300
4. An acute pellet apex (or tip), ×500
5. Arrangements of diverse wood scraps relatively to the whole (vertical, horizontal, oblique) pellet, ×1000
6. Undigested wood cells, detail, ×3000

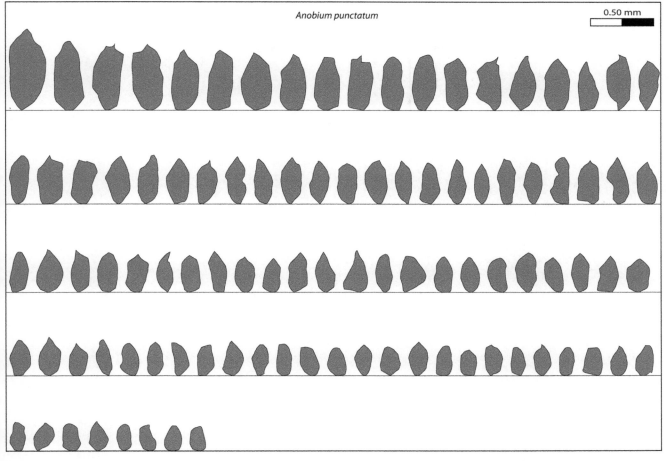

Anobium punctatum

0.50 mm

Drawings representing about 100 fusiform faecal pellets composing *Anobium punctatum* frass. Arranged from the largest to the smallest pellet, collected from the fir sample mentioned above.

Calymmaderus solidus (Keisenwetter, 1877)

Woodworm Beetle

Description
This species is 3.5–5.5 mm in length. It has very characteristic antennae (especially the last three that make up the club), making it easily distinguishable from other common species of the Ptinidae family.

In Europe, the genus *Calymmaderus* comprises only two species. One of them is *Calymmaderus oblongus* (Gorham, 1883), which is present only in the Azores archipelago.

Distribution
Originally from America, this species is now acclimatised in a large area of southwestern Europe, from Cantabria to Galicia in Spain to northern Portugal. It is found in France, Italy, Portugal, Spain.

Bioecology
It appears to be an anthropophilic species. Its biology is poorly known due to a lack of bibliographical data, but seems to be close to the biology of the genera *Anobium* and *Oligomerus*. However, this remains to be verified.

Cycle of Development
Minimum 1 year.

Infested Woods
Not well known. The infested woods studied in this book were found in an interior flooring made of oak (*Quercus*) located in Bayonne (France).

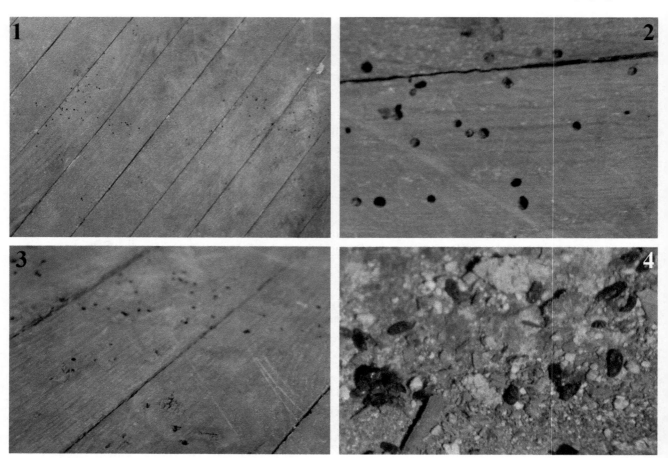

1. View of damage caused by *C. solidus* on a modern wood floor
2–3. Exit holes
4. Dead adult insects, under the floor

The galleries of *Calymmaderus solidus* are round and quite regular, and 1.5–3 mm maximum in diameter. The larvae do not target specific cells. Like the galleries of *Anobium*, the galleries of *Calymmaderus* are preferentially bored in the same direction as the grain of the wood but easily accommodate to obstacles (knots, cracks, varnish, paint …).

Because of their size and circular shape, the adults' flight holes can be mistaken for those of other Ptinidae. The distinction of these species is then made based on the macroscopic and microscopic characteristics of the frass.

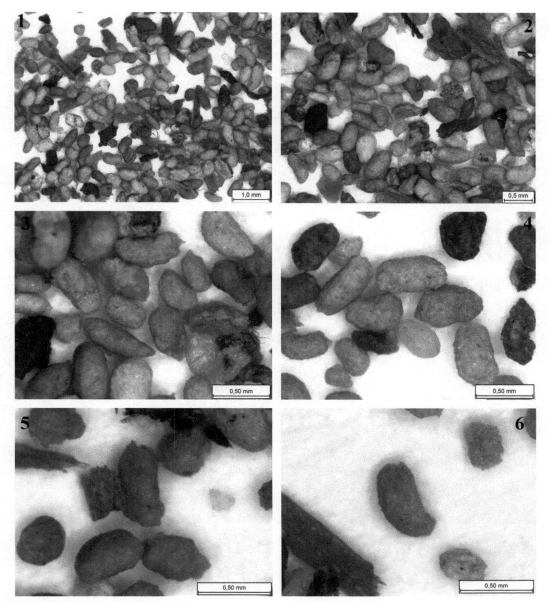

1. Homogeneous frass of *C. solidus,* general view
2–3. Diversity of fusiform pellets: mostly curved, but also sometimes elongated or domed, detail
4–6. Slightly curved pellets

The frass of *Calymmaderus solidus* is homogeneous and granular, and consists of a multitude of fusiform, peanut-like pellets.

Their size depends on the larval stage. Each faecal pellet is about 0.58 mm long by 0.29 mm wide on average. There is no noticeable difference in frass depending on the infested woody species.

The frass is exclusively composed of rather well digested wood. SEM photographs illustrate the sometimes anarchic but compact arrangement of wood scraps within a fusiform pellet.

The pellets are variable in shape, but most of them have an acute apex whose base is curved as compared to the axis of the pellet. Each pellet is compact and difficult to break.

C. solidus frass can be mistaken for *Anobium punctatum*, *Oligomerus ptilinoides* and *Nicobium castaneum* frass. It is necessary to refer to the average measurements and the frequency of details to differentiate between them. Thus, in the Ptinidae group, *C. solidus* forms the most compact faecal pellet and on average the largest ones as compared to other species. The closest species is *Oligomerus ptilinoides*, which can be differentiated based on the domed but uncurved pellets, and by the microscopic arrangement of scraps which are not in a multi-layer in *Calymmaderus* frass.

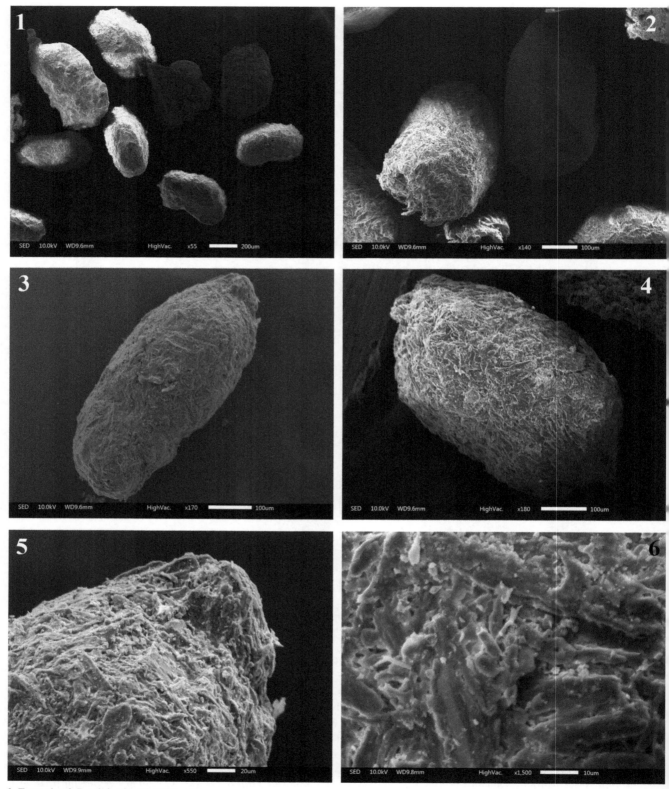

1. Example of *C. solidus* faecal pellet, ×55
2. One fusiform pellet and one calcite stone also found in the gallery, ×140
3. One pellet, general view, ×170
4. Woody scraps on a pellet, detail and arrangement, ×180
5. Acute apex (or tip), detail, ×550
6. Undigested wood cells, detail, ×1500

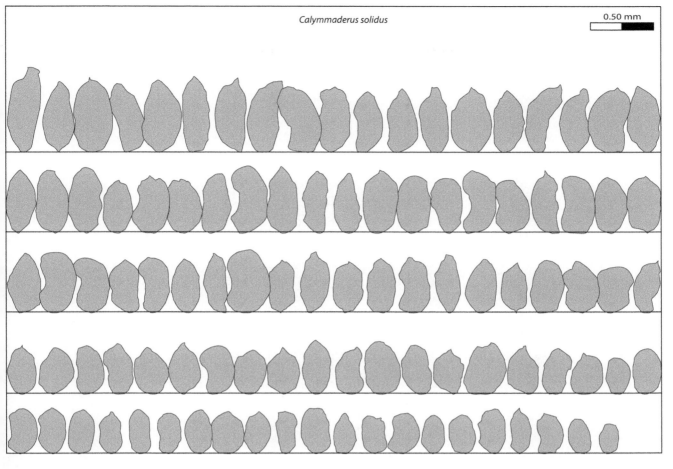

Drawings representing the fusiform faecal pellets composing the frass of *Calymmaderus solidus*. Arranged from the largest to the smallest pellet, collected on an oak (*Quercus*) sample.

Ernobius mollis (Linnaeus, 1758)

Bark Waney Edge Borer

Description
Brown to light reddish (sometimes black), the elongated, subcylindrical imago is 3–6.2 mm long, and covered with a thin, blond, recumbent pubescence.

Like *Xestobium rufovillosum*, this species has unstriated elytra.

Both females and males have large eyes, but males have particularly large and prominent ones. The articles of their antennal club are slightly more elongated than those of the females. They can also be differentiated based on the last tergite of their abdomen, which is slightly raised in males while it is turned downwards in females.

Distribution
Among the species of the genus *Ernobius*, this is the most common and best known one. It is widely distributed in the palearctic and holarctic regions. It was introduced into South Africa, Australia and New Zealand later and is now becoming cosmopolitan. In Europe, it is very present in the pine forests of Scandinavian countries as well as in Great Britain and Ireland.

Bioecology
In the natural environment, the emergence of adults occurs from June to August, with a maximum in July.

After pupation, the adults remain in the wood for about 6–12 days before emerging. After that, they remain mated for about 24 h (up to 6 days), during which they often hide in bark crevices or old flight holes. Adults do not feed, and live for about 1 month.

Eggs (15–25 on average *per* female) are laid singly or in small batches of 2–3, with a maximum of six in the same place. Oviposition mainly takes place around the base of branches, at nodes, in cracks in the bark at the level of protruding, rough bark scales, and in the immediate vicinity of resin canals.

When larvae hatch, they eat part of the shell and usually feed on its remains after emergence. The first-instar larva can move fairly rapidly on the bark and may wander for 1 or more days, covering a distance of several centimetres on the host.

Cycle of Development
The complete cycle lasts about 1 year. Depending on the thermo-hygrometric conditions, it is not uncommon for this cycle to last 2–3 years or more. Some authors indicate that there can be two generations *per* year. This insect can infest the same wood over several generations (from two to six generations), so infestation can occur over several years.

Infested Woods
This species is typically cambio-xylemophagous. It grows in the cambium as well as in the superficial parts of the sapwood of conifers only.

Usually dry wood is infested, but the bark of freshly felled wood and sometimes even physiologically altered and dying trees can also be infested.

E. mollis can be found in manufactured structures as long as they have not been stripped of their bark (roof trusses, beams, boards, pallets, joinery products, sawmills). Even small pieces of bark can cause infestation by this pest.

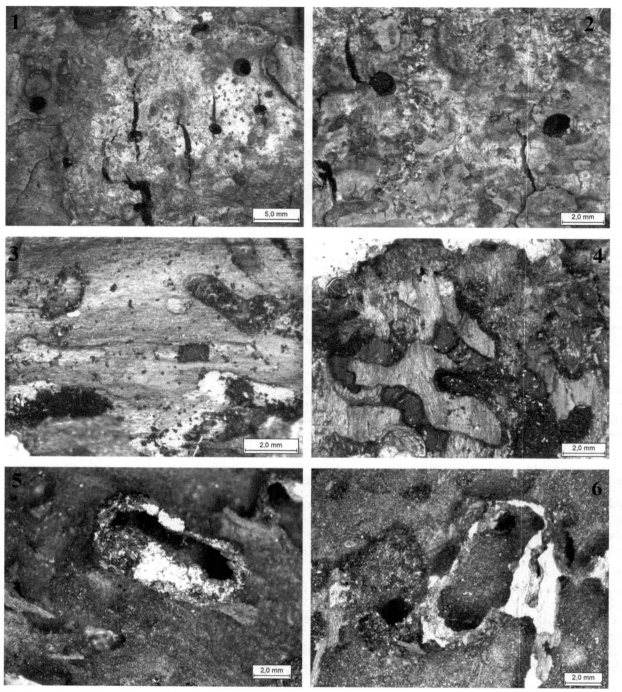

1–2. Circular exit holes of *E. mollis* on cypress (*Cupressus*) bark
3–4. Sinuous larval galleries
5. Fine pupation shell composed in particular of frass, fixed below of the bark
6. Pupation shell with visible exit hole

The galleries of *Ernobius mollis* are round and quite regular, and 1.5–3.5 mm maximum in diameter (usually 2 mm). They are bored mainly parallel to the bark surface on 6–7 cm. The larvae do not target specific cells in the sapwood. The larval galleries are preferentially bored in the same direction as the grain of the wood and become perpendicular to the axis of the tree trunk (or branch) at exit holes.

Because of their size and circular shape, the flight holes of adults can be mistaken for those of other Ptinidae, and sometimes even with those of some Scolytinae (exit galleries only, but no graphiosis along with *Ernobius*). The distinction is then made on the basis of frass.

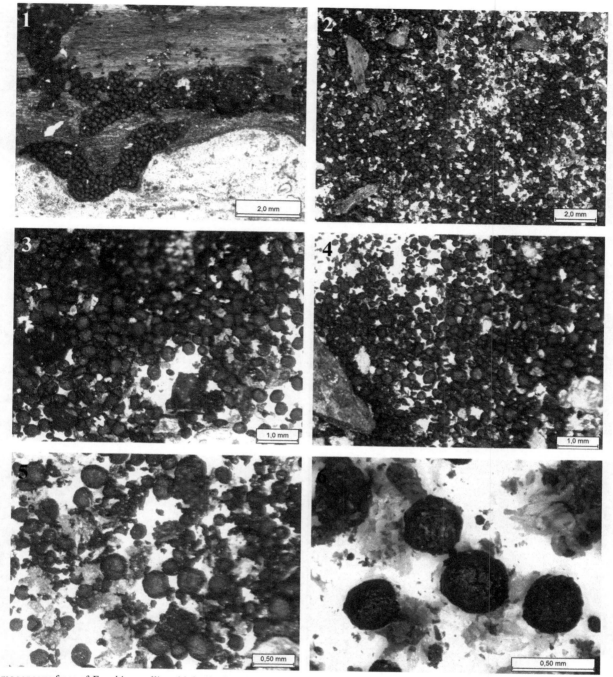

1. Homogeneous frass of *Ernobius mollis*, with lenticular pellets
2. Packed frass in the galleries
3–4. Lenticular pellets of bark colour
5–6. Lenticular pellet, detail

The frass of *Ernobius mollis* is homogeneous and granular, and consists of a multitude of spherical to lenticular faecal pellets, sometimes more irregularly shaped. It is very similar to the frass of *Xestobium*, but much more spherical than lenticular and of a much smaller diameter: 0.35 mm on average for *Ernobius* pellet *versus* 0.58 mm for *Xestobium* pellet. The general colour of the pellets can be that of the sapwood (white, beige grains) and that of the bark (darker grains, brown to black).

Faecal pellets are exclusively composed of rather well digested and well agglomerated wood. SEM photographs illustrate the sometimes anarchic but compact arrangement of wood scraps within a spherical pellet. A suture-like or strip-like stripe can be observed on each pellet, usually on one of the flattened sides of the pellet.

Old frass has eroded pellets, and calcite may be observed.

1. Frass of *E. mollis,* ×200
2. Lenticular pellets, ×250
3. Striations (sutures) on each faecal pellet, detail, ×500
4. Two pellets, detail, ×340
5. Non-eroded pellet, detail, ×800
6. Eroded pellet with calcite formation, detail, ×800

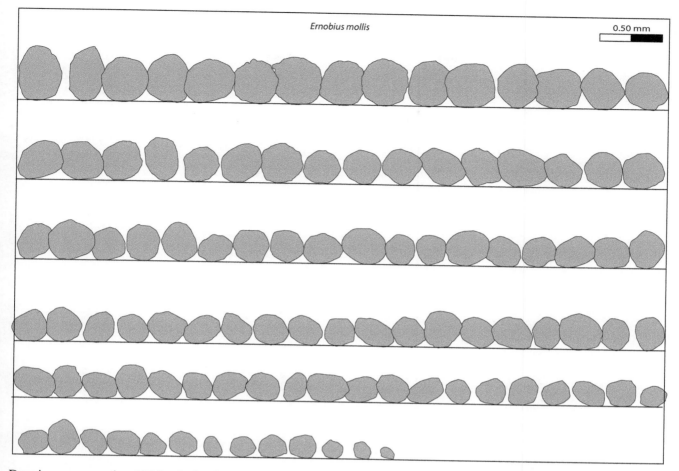

Ernobius mollis

0.50 mm

Drawings representing 100 lenticular faecal pellet or so composing *Ernobius mollis* frass. Arranged from the largest to the smallest pellet, collected on a Cypress (*Cupressus*) sample.

Nicobium castaneum (Olivier, 1790)

Library Beetle

Description

Nicobium castaneum is reddish brown with numerous bristles all over its body. It has strongly convex elytra. It is 4–6 mm long. Like *A. punctatum*, it has a fairly prominent cap but no median elevation. Its elytra are striated with rows of deep dots and have a double—lying and erect—pubescence; this criterion is used to differentiate it from species close to the genera *Anobium* and *Oligomerus*.

The larva is arched, 5–7 mm long and up to 3.5 mm wide.

Distribution

N. castaneum is less common than *A. punctatum*. It typically has a holarctic distribution and remains absent from tropical and South American regions.

It is found in Austria, the Azores Islands, the Balearic Islands, Bosnia and Herzegovina, the Canary Islands, Croatia, Cyprus, the Czech Republic, France, Germany, Greece, Italy, Madeira Island, Malta, Portugal, Romania, Sardinia, Sicily, Slovenia, Spain, Switzerland, Ukraine.

Bioecology

The biology of this species is close to that of *A. punctatum*.

N. castaneum can be seen from early spring to September, with an emergence peak in July and August.

Females lay about 30 eggs on average in several batches, in cracks or old wood galleries. The eggs incubate for about 1 month.

At the end of their development, the larvae build a pupal shell from their excrement, close to the surface of the wood. This shell, which is very resistant, often remains in place long after the adult has flown away and is an identification feature of this species.

Development is optimal when temperatures are around 22 °C, with relative humidity between 55% and 60%. Adults live about 1 month, sometimes more.

Cycle of Development

The duration of larval development is highly dependent on the nutritional quality of the wood, the temperature and the ambient humidity. Thus, the complete cycle of this species can last between 8 and 36 months, and up to 10 years if conditions are unfavourable.

Infested Woods

The library beetle prefers white, soft hardwoods, but can also feed on coniferous woods if necessary, as well as all kinds of wrought and shaped woods. Harder woods such as oak can be infested if a lignivorous fungus has softened and pre-degraded the wood surface. Finally, this species also feeds on old paper or cardboard documents.

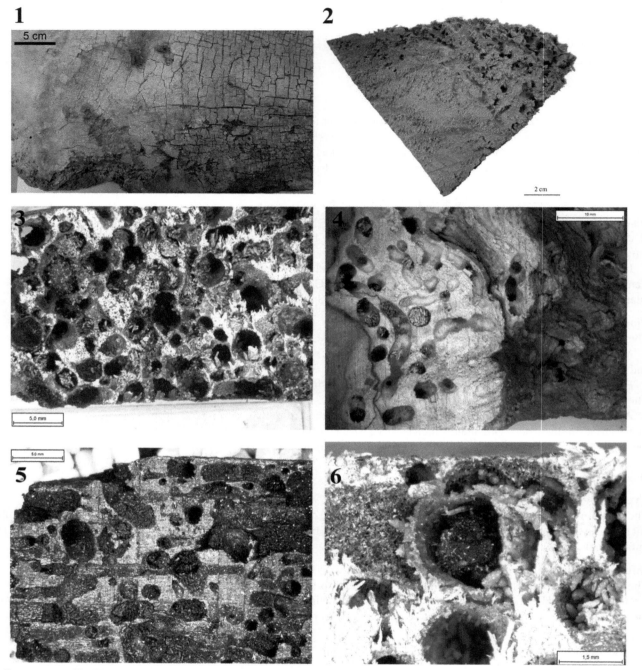

1. Damage caused by *N. castaneum* on to an oak (*Quercus*) beam infested by cubic rot, general view
2. Damage caused by *N. castaneum* on poplar (*Populus*), general view
3. Galleries of *N. castaneum* on agglomerated wood, high infestation
4. Galleries in oak (*Quercus*) sapwood previously infested by a lignivorous fungus
5. Galleries of *N. castaneum* on experimentally carbonised agglomerated wood, high infestation
6. Pupal shell with imago blocked inside

The galleries of *N. castaneum* are round and quite regular, and 1.5–3.5 mm maximum in diameter (2.3 mm on average). The larvae do not target specific wood cells and generally bore in the same direction as the grain of the wood. On the other hand, they are often found along with lignivorous fungi such as cubic rot fungi, and seem to follow the same progress in the wood. When the generations follow one another on the same object or host, the larval galleries intersect.

The galleries can be mistaken for those of *Anobium punctatum*, *Calymmaderus solidus*, and *Oligomerus ptilinoides*, and these three species are differentiated on the basis of frass. This species is more easily spotted by the naked eye than other dry wood Ptinidae because of its typical pupal shells, usually located on the wood surface.

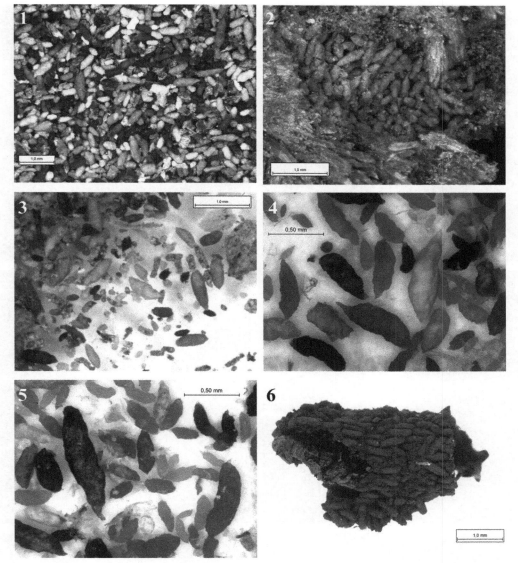

1. Frass of *Nicobium castaneum*
2. Fusiform faecal pellet in poplar (*Populus*) gallery
3–4. Fusiform pellets, detail
5. Example of *N. castaneum* frass on paper
6. Pupation shell built by the larva with fusiform pellets and pupae inside, experimentally carbonized oak sample

The frass of *N. castaneum* is homogeneous and granular. It consists of a multitude of fusiform faecal pellets that look like elongated peanuts.

The size of the pellets depends on the larval stage. Each faecal pellet is about 0.57 mm long by 0.21 mm wide on average. There is no noticeable difference in frass depending on the infested woody species. On the other hand, a difference in texture is observed when the frass is from a paper support (old books) that *Nicobium* is fond of. The pellets then have a more shrivelled, wrinkled and fragile aspect.

The frass is exclusively composed of digested wood. SEM images show a hollow line (or stripe) along the entire length of the pellet.

The details of the pellets are variable, the ends of a fusiform pellet are most often sharp.

Nicobium castaneum is the only species of Ptinidae on dry wood that produces a pupal shell built exclusively with fusiform pellets arranged mainly lengthwise.

The frass can be mistaken for those of *Oligomerus ptilinoides*, *Calymmaderus solidus* and *Anobium punctatum* but to differentiate between them one can refer to average measurements and frequency of detail. Thus, in the Ptinidae group, *N. castaneum* produces the most elongated and best digested faecal pellet of all other species.

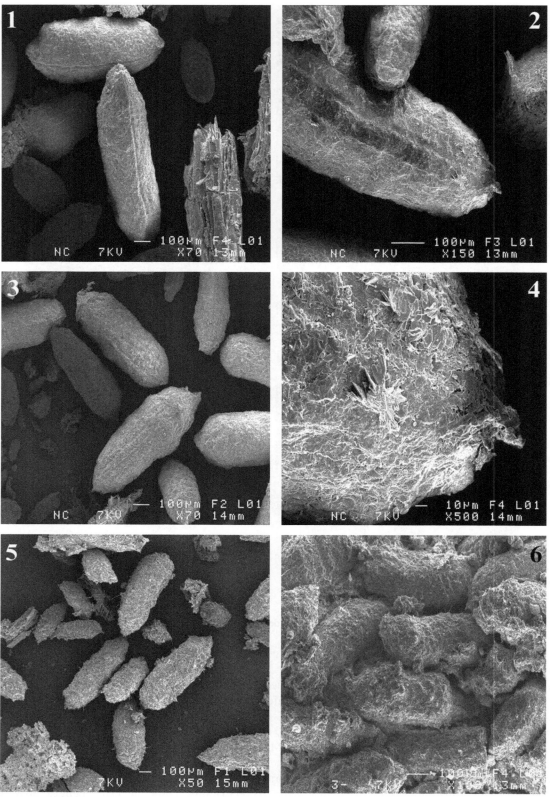

1. Faecal pellet of *N. castaneum*, ×70
2. Pellet with a visible stripe, detail, ×150
3. Homogeneous frass of *N. castaneum*, ×70
4. Acute apex (or tip) with well-digested wood arrangement, detail, ×500
5. Frass of *N. castaneum*, carbonized, ×50
6. Faecal pellet of *N. castaneum*, composing a pupation shell, carbonized, ×100

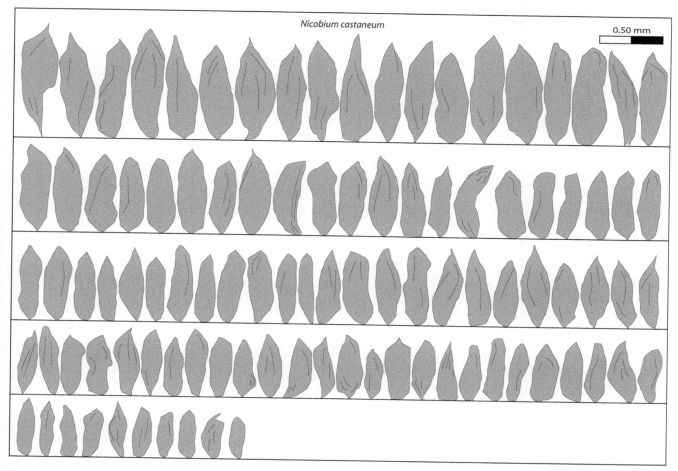

Drawings representing 100 fusiform faecal pellets or so composing the frass of *Nicobium castaneum*. Arranged from the largest to the smallest pellet, collected from the composite wood sample.

Oligomerus ptilinoides (Wollaston, 1854)

Furniture Beetle Woodworm

Description

Oligomerus ptilinoides is brown to reddish-brown and has a strongly rounded pronotum that almost completely covers its head, which is usually a little darker than its elytra. This species is 5–7 mm long. Its elytra are regularly streaked with lines of faint dots and covered with a short down of fine light bristles bent backwards. Its antennae consist of 11 items, the last three of which are much more elongated than the others.

The larva is arched and is 6–7 mm long and up to 3.5 mm wide.

Distribution

The species is present throughout the Mediterranean basin (south of France, Spain, North Africa). It is absent from tropical regions.

It is found in Austria, the Balearic islands, the Canary islands, Croatia, Cyprus, France, Germany, Greece, Hungary, Italy, Madeira island, Malta, Poland, Portugal, Romania, Sardinia, Sicily, Slovakia, Slovenia, Spain, Switzerland, Ukraine.

Bioecology

Its biology is very close to that of *A. punctatum* and *N. castaneum,* except that *Oligomerus* is strictly xylophagous. The other two species can and do develop at the expense of more diversified materials such as cardboard and paper from books and archives. The optimal conditions for the development of *O. ptilinoides* are identical to those mentioned above except that its maximum temperature for larval development is a little higher and can reach 32 °C. This species is therefore considered thermophilic and that is why its range is in the Mediterranean area.

When the ambient temperature is below 12 °C, the larva can fast for about 3 months.

As with *A. punctatum,* the larva does not build a pupal shell but occupies a space below the wood surface.

Cycle of Development

The development cycle varies from 1 to 3 years depending on thermo-hygrometric conditions, the nutritional quality of the wood and the presence of lignivorous fungi. It is approximately 1 year in Mediterranean regions and 2–3 years in areas with a more temperate climate.

Infested Woods

This insect infests various materials: furniture, statues, works of art, carpentry …

It prefers deciduous wood species (elm, ash, oak, maple, apple …).

It causes proportionally more damage than *Anobium punctatum* because the adult insect is larger.

Oligomerus brunneus, a very rare species of this genus, can also be found on timber. At present, however, very little is known about its lifestyle and biology.

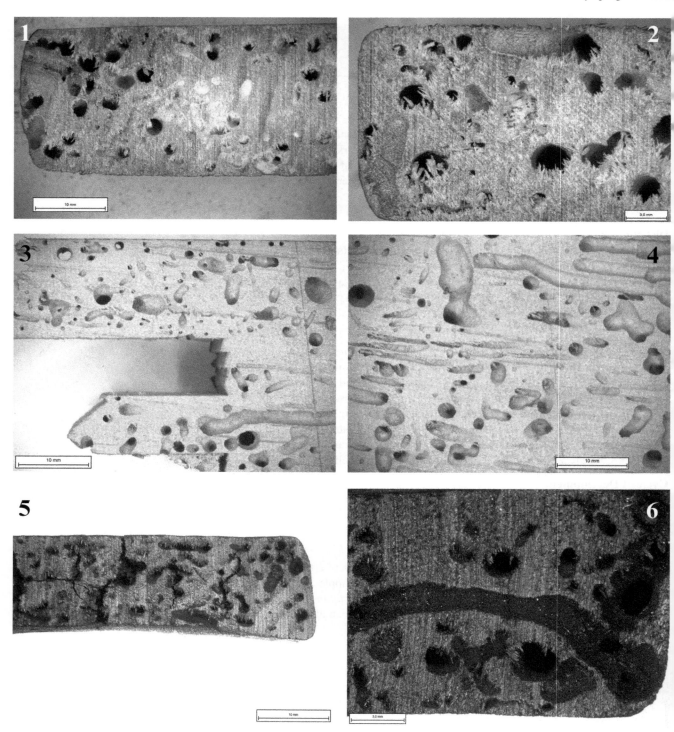

1. Galleries of *Oligomerus ptilinoides* in a lime tree (*Tilia*) sample
2. Zoom of exit holes of *Oligomerus ptilinoides*
3–4. Damage caused by *O. ptilinoides* to beech, general view
5–6. Experimentally carbonized galleries

The galleries of *O. ptilinoides* are round and quite regular, with a diameter varying from 1.5 to 3.5 mm maximum (2.8 mm on average). The larvae do not target specific wood cells and generally bore in the same direction as the grain of the wood. When the generations follow one another on the same object or host, the larval galleries intersect.

The galleries can be mistaken for those of *Anobium punctatum*, *Calymmaderus solidus* and *Nicobium castaneum*, although they have a greater average diameter. These three species are differentiated on the basis of the frass.

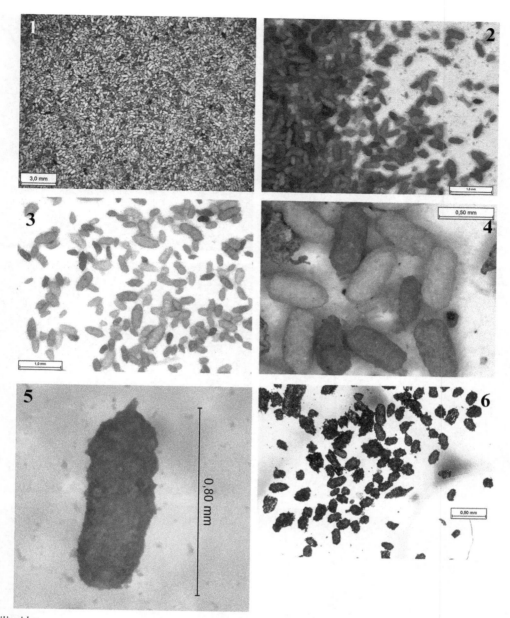

1. Frass of *O. ptilinoides*
2–3. Frass and fusiform faecal pellet
4–5. Domed pellets, detail
6. Domed fusiform faecal pellet in an archaeological charred beech (*Fagus*) sample, Roman period (Fréjus, France)

The frass of *O. ptilinoides* is homogeneous and granular. It consists of a multitude of fusiform faecal pellets that look like bulging peanuts.

Their size depends on the larval stage. A larva is about 0.61 mm long and 0.31 mm wide. There is no noticeable difference in frass depending on the infested woody species. Each faecal pellet is exclusively composed of more or less digested wood. SEM photographs illustrate a parallel and multi-layered arrangement of the various wood scraps.

The base of the top of the pellet is generally rounded and domed and ends in a pointed apex. The other end is often rounded. Pellets are stable, and do not settle or crumble except in rare cases.

The frass of *O. ptilinoides* can be mistaken for the frass of *Anobium punctatum, Calymmaderus solidus* and *Nicobium castaneum*. One can refer to the average measurements and the frequency of details to differentiate between them. Thus, in the Ptinidae group, *O. ptilinoides* produces the most domed faecal pellet with a parallel arrangement of wood scraps, whereas the other species do not. The closest species is *Calymmaderus solidus*, but the curvature of its faecal pellets is not found in *Oligomerus,* and digested wood scraps are arranged in a multi-layer in its pellets *versus* a rather anarchic arrangement in *Calymmaderus* pellets.

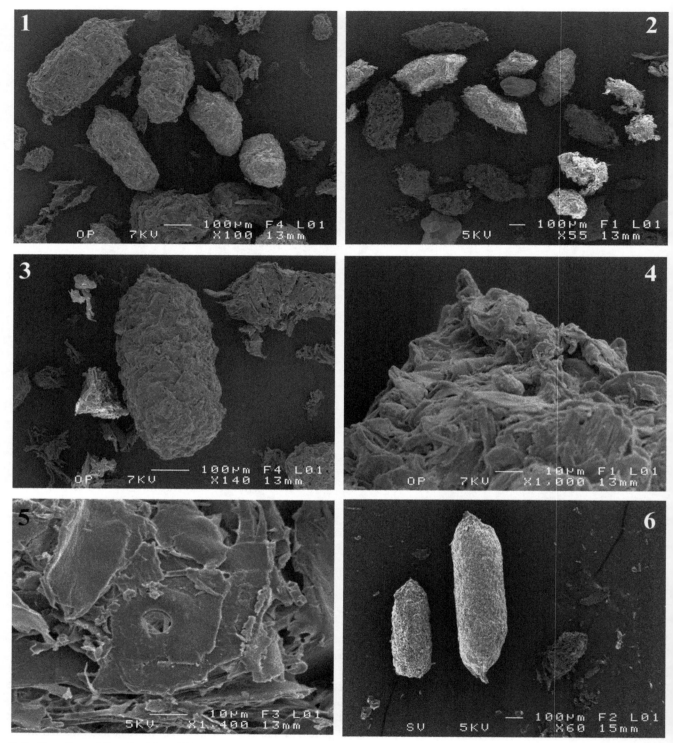

1. Fusiform frass of *O. ptilinoides*, ×100
2. Fusiform pellets, charred, ×55
3. Fusiform and domed pellet, with parallel arrangement of more or less digested wood scraps, ×140
4. Acute apex (or tip) with digested wood arrangement, ×1000
5. Carbonised, undigested wood cell, ×1400
6. Comparison of the fusiform pellets of three species of Ptinidae. From left to right: *O. ptilinoides, N. castaneum, A. punctatum*, ×60

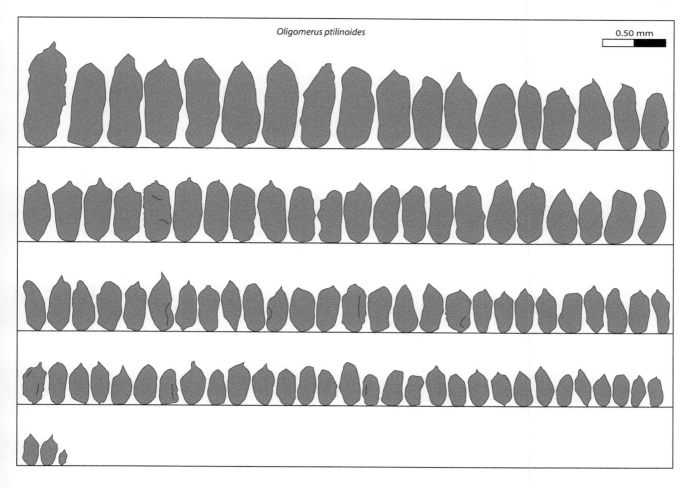

Oligomerus ptilinoides

0.50 mm

Drawings representing 100 fusiform faecal pellets or so composing the frass of *Oligomerus ptilinoides*. Arranged from the largest to the smallest pellet, collected from a lime tree (*Tilia*) sample.

Xestobium rufovillosum (De Geer, 1854)

Deathwatch Beetle

Description
Reddish brown, *Xestobium rufovillosum* is 5–8 mm long.

This species is relatively stocky and has a large, broad, triangular thorax, rounded at the corners. Its elytra are finely granulated with irregular tufts of short, flat, golden bristles that give it a marbled aspect. Its antennae consist of 11 articles, the last three of which form a weakly developed club. The arched larva is 10–11 mm long and up to 4.5 mm wide.

Distribution
Cosmopolitan, *X. rufovillosum* is well represented in Europe and is also present in North America and Australia.

Bioecology
This species is among the largest ones in this family and therefore causes much more damage than *Anobium punctatum*, even though they are both very common and evolve in the same ecological biotopes.

Adults emerge much earlier in spring than the other species of this family, starting in April. The eggs (60–100 on average) are laid in small batches, preferably in cracks and old wood galleries, and incubate for about 15 days.

This insect develops in wood previously infested by lignivorous fungi, and the nutrients needed by the larvae are supplied by the nitrogen released by these fungi (responsible for cubic or fibrous rot).

This species needs a minimum temperature of 22–25 °C and a relative humidity of 22% to establish in the wood and develop.

The adult has an average longevity of 8–10 weeks, sometimes much longer.

An increase in infestations by this species can be observed after certain particular climatic events (heavy rains, floods, typhoons …).

Cycle of Development
At least 2 years (in optimal conditions), 3 on average, and exceptionally up to 10 years.

Infested Woods
X. rufovillosum is one of the wood-boring insects that causes the most damage to old carpentry, woodwork, floors, and any wooden structure after *Hylotrupes bajulus*.

This species prefers the sapwood of deciduous trees: oak (*Quercus*), birch (*Betula*), alder (*Alnus*), elm (*Ulmus*), willow (*Salix*), chestnut (*Castanea*) and beech (*Fagus*). Tropical woods are generally little infested or uninfested.

1. Damage caused by *X. rufovillosum* on oak (*Quercus*) planks
2. Larval galleries, detail
3. Thin slide of an oak (*Quercus*) sample infested by *X. rufovillosum*, cl. A. Crivellaro

The galleries of *X. rufovillosum* are round and quite regular, with a diameter varying from 2 mm for the early larval stages to 4.5 mm for the exit hole (3.5–4 mm on average).

The larvae do not target specific wood cells and generally bore in the same direction as the wood grain. On the other hand, they are often found along with lignivorous fungi such as cubic rot fungi and seem to follow their progress in the wood. When the generations follow one another on a same object or host, the larval galleries intersect.

The galleries can be mistaken for those of other Ptinidae, Bostrichidae and Lyctidae based on their general shape, but the much larger size of the exit galleries rules out the hypothesis of the presence of other species. Other insects bore galleries of the same size or even larger, but do not generally produce frass.

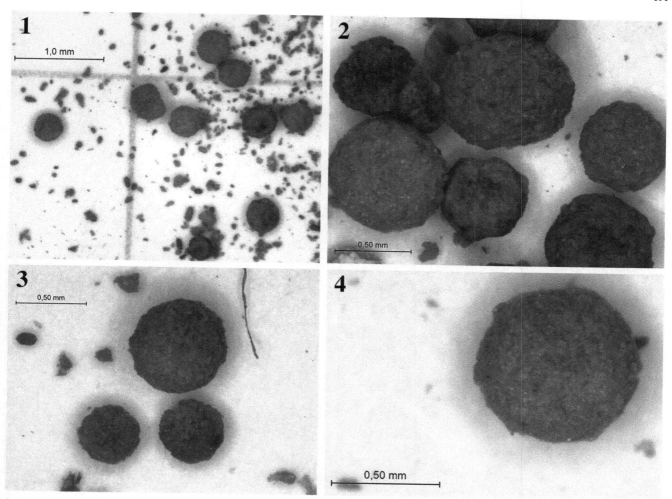

1. Homogeneous and granular frass of *X. rufovillosum*
2–4. Faecal pellet, typical spherical to lenticular shape

The frass of *X. rufovillosum* is homogeneous and granular. It consists of a multitude of flattened, spherical, lentil-shaped pellets sometimes mixed with various wood scraps coming from the packing of the frass.

Each faecal pellet measures 0.6 mm in diameter on average, with a maximum of about 1 mm.

The frass of *X. rufovillosum* is very similar in shape to the frass of *Ernobius mollis*, but the size and location of the galleries in the wood leave no doubt as to its identification.

The faecal pellet is exclusively composed of more or less digested wood. SEM photographs illustrate the random arrangement of the various wood scraps within a pellet, which makes it brittle. There are no visible striae, unlike in *Ernobius* frass.

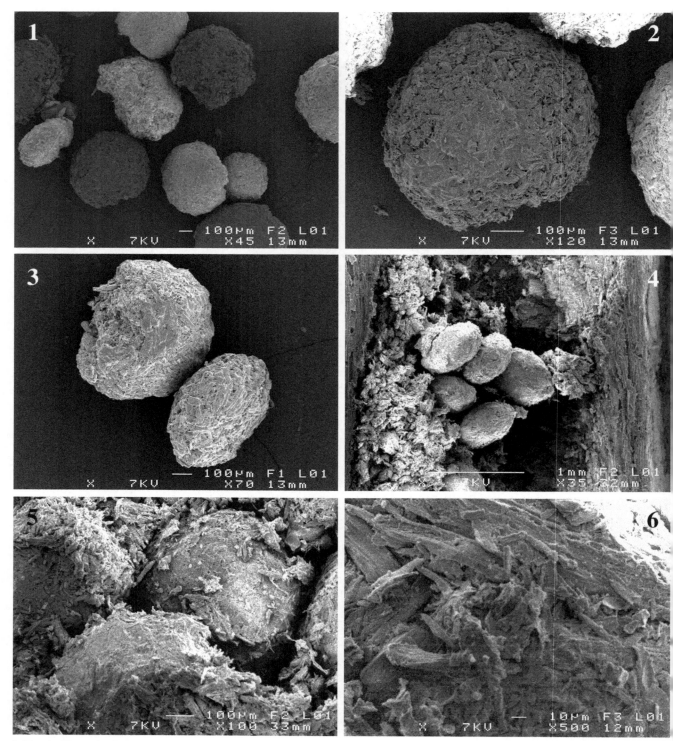

1. *X. rufovillosum* lenticular pellet, ×45
2. Lenticular pellets, and wood fragment arrangement ×120
3. *X. rufovillosum* lenticular pellet, side view, ×70
4. Lenticular pellet with gallery packing, ×35
5. Packed pellets, detail, ×100
6. Arrangement of wood scraps in a pellet. No wood cell observed, ×500

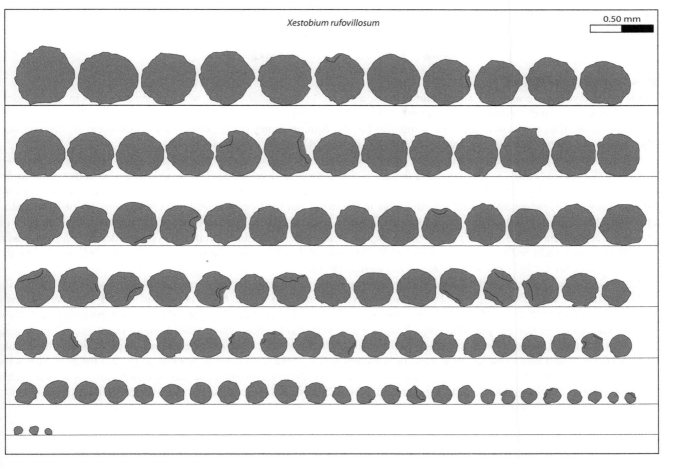

Drawings representing 100 fusiform faecal pellets or so composing the frass of *Xestobium rufovillosum*. Arranged from the largest to the smallest pellet, collected from the oak (*Quercus*) sample mentioned above.

3.2   Hymenoptera

3.2.1   *Apidae*

Presentation of the Family

In the order Hymenoptera (bees, wasps, bumblebees, ants, etc.), the Apidae family, which represents bees in the broadest sense, forms the most diversified group with more than 5700 species.

Three subfamilies can be distinguished within this family: Apinae (honey bees), Nomadinae (cleptoparasitic insects), and Xylocopinae or carpenter bees, one species of which is presented here: *Xylocopa violacea.*

The subfamily Xylocopinae comprises four tribes, including Xylocopini, which includes only one homogeneous genus—*Xylocopa* (Latreille, 1802)—whose common feature is that it nests in dead wood. This is indeed the meaning of the word *Xylocopa* established by Latreille in 1802 (*xylos*: wood, and *koptein*: to cut). In fact, it is not a real xylophagous insect (with larvae eating wood) but a nesting one: the adult digs in the wood, but the larva does not (except to come out of its nymphal chamber). Most species of this genus look like a bumblebee, sometimes hairy, with a black body and wings with metallic or purple reflections about 2–3 cm long. Xylocopes include seven species in Europe (eight if we consider *X. uclesiensis* as a species in its own right and not as a subspecies of *X. iris*), e.g., *Xylocopa valga* (Gerstaecker, 1872), *X. violacea* (Linnaeus, 1758), *X. iris* (Christ, 1791) and *X. cantabrita* (Lepeletier, 1841).

Carpenter bees, or xylocopic bees, are solitary nesting insects that settle in the wood, like certain species of Megachile and some ants do. They dig a long tunnel in the wood and segment it into several boxes closed by chewed wood scraps. They deposit one egg in each box and a pollen agglomerate to feed the larvae. Females generally prefer unpainted wood. New females can reuse old tunnels year after year. They are also attracted to areas where other females dig tunnels. Egg-laying and provisioning of the galleries takes place in the spring. The males circle around the gallery entrance while the female supplies the nest and lays eggs. Because these bees are not social, there is no caste of workers to protect the nest. Stings by females are rare, while males do not sting. The new adults emerge after mid-summer and can be seen feeding on flower nectar until they search for wintering sites. Over generations, the galleries can compromise the mechanical strength of the wood.

Xylocopa violacea (De Geer, 1774)

Large Carpenter Bee, Purple Xylocopa

Description

The purple xylocope is 18–28 mm long (mostly 20–25). It resembles bumble bees in size and shape, and its overall colour is black with hairs of the same colour for the female. Its wings have a metallic blue-black to purplish metallic sheen, hence their name. The male has grey shades with a hairy thorax corresponding to a mixture of black and light-coloured bristles. Compared with the female, it also has two articles with a reddish-pink apex.

This species differs from the related species *Xylocopa valga* (Gerstäcker, 1872) by different external morphological features (Terzo 2007).

Distribution

It is present throughout the Mediterranean area as it is a thermophilic species that seeks out sunny places.

Bioecology

The first imagoes come out in February-March. A second generation may come out in October.

As soon as temperatures become milder, the female digs a gallery in pre-degraded wood (degraded by a fungus, for example). To do this, it detaches various wood scraps that she rejects outside the wood. Therefore, there is no faecal pellet as such, but rather wood scraps. She builds a batch of cells (or boxes), sometimes superimposed ones, using chewed wooden partitions, and lays one egg in each them, together with an agglomerate of pollen to feed the young larvae. The larvae hatch a few days after egg-laying and then take 3 weeks to develop, at the end of which they turn into nymphs. Emergence will only take place after about a month. The larvae either perforate the cell walls or wait for the upper neighbour to leave.

Adults spend the winter in the imago state and take shelter in the holes of old walls or in the cavities of hollow trees. They tend to return regularly to the pieces of wood from which they came, and thus carry on with the infestation.

Cycle of Development

The complete cycle varies from one and a half to 2 months. Each female builds 1, 2 or 3 nests throughout a season. Adults live about 1 year.

Infested Woods

Both deciduous and coniferous trees can be taken as hosts by the carpenter bee. Females nest in all kinds of dead wood such as tree trunks or large branches, but also in posts, palisades, beams, etc. Carpentry wood can be seriously damaged by repeated nesting over several years.

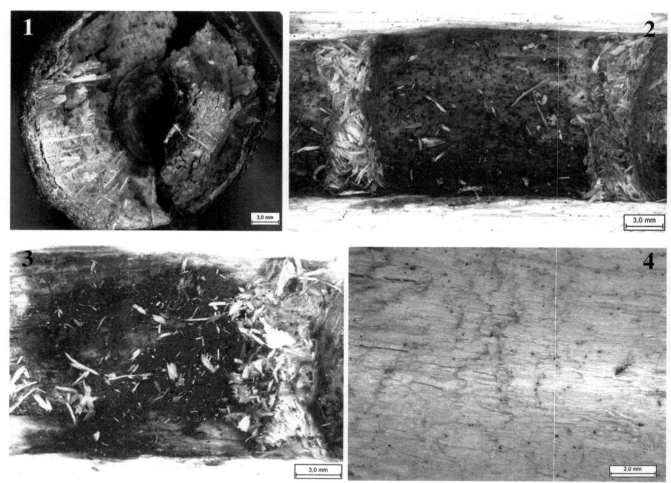

1. View of a gallery of *Xylocopa violacea* in a branch of oak (*Quercus*) stored under shelter

2–3. Longitudinally observed gallery with compartments and walls

4. Bottom of a compartment with traces of the adult tearing off wood scraps during gallery excavation

The adult female digs round and regular galleries 10–15 mm in diameter. The adult generally extends them along the grain of the wood over about 10 cm or even up to more than 40 cm if the support allows so.

The gallery is divided into several compartments separated by walls made up of scraps chewed by the bee. The exit of the gallery is plugged by the female bee in the same way.

The galleries are probably similar to those dug by other xylocopes.

1–3. Wood scraps forming the walls of the nymphal boxes, detail
4. Pollen agglomerate in the bottom of a lodge

As the wood is not ingested during excavation but rejected in the form of coarse sawdust, it does not form faecal pellets as such. However, various wood scraps can be found: long, coarse shavings just under 1 cm in length making up the walls of the lodges; and pollen agglomerates deposited in each individual compartment.

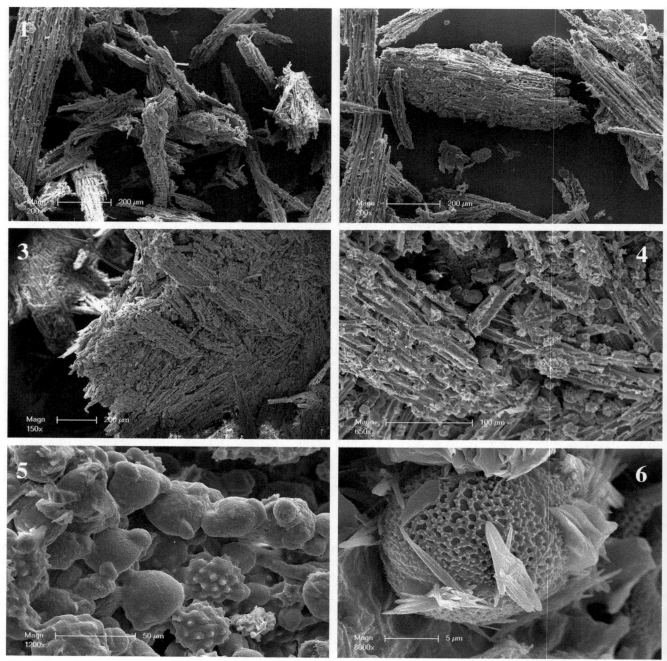

1. Wood scraps found in a gallery of *X. violacea*, ×200
2. Wood scraps, ×200
3. Chewed wood scraps, nymphal lodge wall scrap, ×150
4. Pollen stuck on a wall, ×650
5. Pollen agglomerated at the bottom of the nymphal lodge, detail, ×1200
6. Pollen grain, ×8000

3.2.2 *Formicidae*

Presentation of the Family

In the order Hymenoptera (bees, wasps, bumblebees, etc.), the Formicidae family, commonly known as ants, includes about 400 European species. Like bees, these insects are social and form colonies of a few dozen to several million individuals divided into castes (workers, queens, and males in particular), each of which plays a well-defined role within the colony. In Europe, this family is composed of seven subfamilies. The present atlas deals with the so-called carpenter ants. The genus described here is *Camponotus* (subfamily Formicinae, tribe Camponotini). It includes several species nesting in old stumps, dead trunks and branches as well as in the often damp wood of houses. It should be noted that many other genera of this family nest in wood (*Crematogaster*, *Lasius*, *Formica*, *Tapinoma*).

Among the species most frequently found in wood in human dwellings are:

- *Camponotus herculeanus* (Linnaeus, 1758): it is the largest in size and is black to reddish-brown. Rare occurrences have been reported inside houses. It is most often found in standing trees.
- *Lasius niger* (Linnaeus, 1758) (the brown ant): it prefers damp woods inside houses.
- *Lasius fuliginosus* (Latreille, 1798) known as the black wood ant: it inhabits dead trees, stumps or floorboards damaged by dampness.
- *Crematogaster scutellaris* (Olivier, 1792), recognisable from its red head and heart-shaped stomach: it nests mainly in wood (stumps, trunks, etc.) and is found in the wood of building roofs.

Like carpenter bees, they are non-xylophagous insects that nest in wood. Ants burrow into the wood, whether softwood or hardwood, settle in it and get rid of the scraps outside the wood. Their galleries serve as a pantry, where they preserve rather sweet foods.

The galleries form large cavities with clean, smooth and worm-free walls. The choice to install a colony seems to depend on the moisture content of the wood and the natural penetration routes (cracks, crevices ...). The diameters of the galleries vary between 10 and 15 mm.

Camponotus vagus (Scopoli, 1763)

Carpenter Ant

Description
Camponotus vagus is one of the largest ants in Europe. Workers vary from 6 to 16 mm in size, with great variation within a same colony. Its body is entirely black, with significant white hairs on its stomach that are much less visible in *Camponotus aethiops* (Latreille, 1798).

Distribution
This species is xerophilous. It appreciates mild climates and is present in most European countries bordering the northern part of the Mediterranean basin: Portugal, Spain, France, Italy, Greece, Italy, etc.

Bioecology
This species is quite common and has the same social characteristics as other ants. It forms colonies with a maximum population of 10,000 workers, usually around 3000 individuals. A colony of carpenter ants does not necessarily start in the wood (under a stone for example) but the colony moves quickly through the wood such as a log on the ground, a tree hole, a stump or a structural or floor element. They dig into the wood with their jaws, using them as gouges. Unlike termites, they do not eat wood. It has no nutritional value for them, so they discard it by dropping it out of the nest area or by stacking it in one place. The excavation process of ant-hills results in galleries with very smooth sides. Large colonies can cause structural damage. The most common outdoor habitat is a living tree with a rot spot on the inside. Other common sites are stumps or firewood. Some colonies are brought in with building materials.

Carpenter ants feed on honeydew, insects and ripe fruit juices. Their exploration territory extends over a radius of about 10 m.

Cycle of Development
This species takes place in April-May. At 25 °C, it takes between 6 and 8 weeks to go from egg to imago, and more than 2 months for the biggest workers called "soldiers". The humidity rate should be very low, about 10% on the surface of the nest.

Infested Woods
This species is found in wooded environments, with a preference for coniferous forests where it nests in ground trunks, stumps and dead trunks left standing (candles). Colonies can also invade wooden structures (frames, floors, shelters, etc.) and cause significant damage.

1. Galleries of a colony of *Camponotus vagus* visible from outside the nest of the colony hidden under bark
2. Galleries of a colony visible across a stored tree trunk
3–4. Inside of the galleries of the colony with wood scraps stored in places

The galleries of *Camponotus vagus* are irregular, about 1 cm in diameter, and run in all directions of the grain of the wood. Ants sometimes use the cracks in the wood to drill the network of galleries. Some galleries widen out to form a "room" where eggs or harvested foodstuffs are placed. The galleries are free of worm holes. The walls are smooth and free of mud (unlike those of *Reticulitermes* subterranean termites).

3.3 Isoptera

3.3.1 *Kalotermitidae*

Presentation of the Family

Termites form a very large group that would require a work project or a work of its own both to investigate their biology and to observe and describe their society.

Termites or Isoptera are similar to ants and bees in their complex societal organisation. There are more than 2000 species of termites in the world, the majority of which are of tropical or equatorial origin. Among them, in Europe, two main types of termites can be distinguished and are distributed into 11 species: subterranean termites (*Reticulitermes*) and dry-wood termites represented by Kalotermitidae, which include the recent introduction of the species *Cryptotermes brevis* (modern import).

The Kalotermitidae family is composed of three genera in Europe: *Bifiditermes rogierae, Cryptotermes brevis* and the genus *Kalotermes* which includes two species, one of which is only present in the Azores archipelago. Two new species—*Kalotermes italicus* and *Kalotermes phoenicae*—have been described recently, and other clades have been identified using molecular tools (Scicchitano 2017). This suggests that the overall diversity of the family is underestimated.

Kalotermitidae found their colony in the wood itself, unlike subterranean termites whose colonies are in the soil in wet conditions. The last stages of Kalotermitidae larvae provide for the vital needs of the colony. They form smaller colonies than subterranean termites, with only several hundred individuals, and can live without association or communication with a moisture source. The individuals are content with the moisture content of the wood. Kaloterm species are considered to be "one-piece" nesters, where individuals nest in and feed on the same substrate.

The newly formed breeders "fly" (or rather hover) all together from the colony, fall to the ground, tear off their wings and try to find a sexual partner from a different colony. The principle is to create a new colony and thus ensure genetic mixing. Thus, the only adults in a termite colony are the initiating couple (easily recognisable because they are the only ones to be pigmented).

Almost all wood species can be infested. These termites can establish their colony in dead trees, but also in wooden structural elements, furniture and carpentry, even if the wood is moved from one place to another. They enter wood through cracks, crevices or even through old exit holes of wood-boring larvae (longhorn beetle, *Lyctus*, Ptinidae).

Kalotermes flavicollis (Fabricius, 1793)

Yellow-Necked Dry-Wood Termite

Description
Winged breeders are 8–10 mm long. They are pale yellow to dark brown and have a yellow thorax, hence their name *flavicollis*. They have two pairs of identical, membranous and slightly smoky wings. The nymphs or older immature adults, which perform the functions of underground termite workers, are 4–6 mm long and are white or cream-coloured. Soldiers are 8 mm long, whitish, with a prominent head armed with strong mandibles.

Distribution
This species is found in the Mediterranean basin, namely in Spain, France, Italy, Greece, Slovenia, the Near East and North Africa. It is also found in continental Portugal and it has recently been identified in a tree infestation in the Azores Islands.

Bioecology
This termite remains inconspicuous and hidden in many kinds of wood where it feeds and builds its nest. It is only when swarming occurs in autumn that they can be seen coming out in large numbers. A few hundred eggs are laid each year. The larvae hatch from the eggs and develop until they reach the stage of old nymphs or immature adults called "false workers".

   K. flavicollis is known to have a high potential to become an invasive species, favoured by the fact that it is easily transported by human activities.

   It can indeed be transported to new habitats and regions in both potted trees and wood. Colonies are made up of 500–1000 individuals on average.

   While *Cryptotermes brevis* nests in dry, healthy wood with a moisture content generally below 15%, *K. flavicollis* is much more demanding: it more easily infests wood that has undergone fungal development and has a moisture content of at least 18–20% or more. Its presence therefore often indicates a problem with moisture and fungal development and can be a good indicator of a climate disorder. However, unlike *C. brevis*, this species has a very low capacity to re-infest or start new colonies from the initial point of infestation.

Cycle of Development
Information is variable and scarce.

Infested Woods
This species can infest a wide variety of ornamental and fruit trees such as poplar, elm, cork, pine, olive, chestnut, fig and pear trees. Damage is often punctual and well localised.

   It generally infests living trees from scars and progresses into the heartwood, partially infesting dead branches. It sometimes gnaws at the heartwood and roots of living trees. In homes it can be found in lumber, parquet, panelling, but also in painted frames and various other wooden objects.

1. Breeding of *K. flavicollis* on vine (*Vitis vinifera*)
2. Galleries of a colony of *K. flavicollis* giving the wood a hollow aspect
3–4. Inside of the galleries of a colony

The galleries of *K. flavicollis* are irregular to round, and 1–4 mm in diameter. There does not seem to be any particular direction in relation to the grain of the wood. The wood takes on a leafy aspect as the colony grows. The set of galleries forms a network with punctual enlargements. Walls are smooth.

1–2. Homogeneous frass of *K. flavicollis* consisting of concave pellets shaped like corn kernels
3–4. Faecal pellets, detail

The nymphs (pseudo-workers) dig galleries and leave faecal pellets. These pellets are homogeneous and very peculiarly shaped: they consist of a multitude of smooth units resembling 0.5–1 mm-long corn kernels with concave sides, with no undigested or poorly digested wood elements. This frass should not be mistaken for any other frass in the current state of knowledge, but little is known about the frass of other termite species.

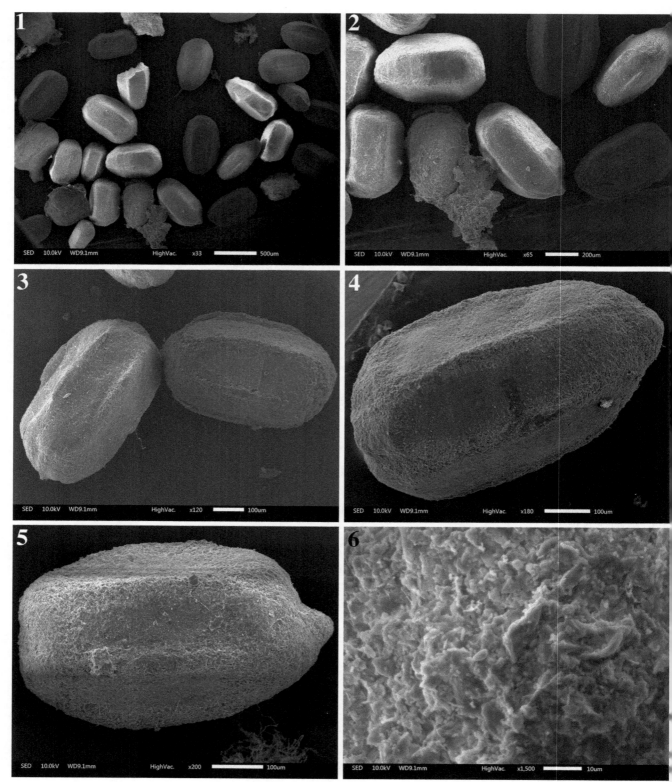

1. Frass of *K. flavicollis* consisting of pellets shaped like corn kernels, ×33

2. Corn-kernel-like pellets, ×65

3. Concave pellets with visible flattened sides, ×120

4–5. One pellet, detail, ×180 to ×200

6. Smooth surface of one pellet without wood cells, detail, ×1500

Index of Species

Latin name	Common name	Subfamily	Family	Order	Page mention
Anobium punctatum	Common furniture beetle	–	Ptinidae	Coleoptera	1, 2, 18, 19, 20, 148, 149, 159–163
Bostrichus capucinus	Capuchin beetle	–	Bostrichidae	Coleoptera	16, 25, 27, 33–36
Callidium coriaceum	Capricorn beetle	–	Cerambycidae	Coleoptera	14, 20, 37, 43, 71–75
Calymmaderus solidus	Woodworm beetle	–	Ptinidae	Coleoptera	18, 20, 148, 149, 165–169
Camponotus vagus	Carpenter ant	–	Formicidae	Hymenoptera	14, 199, 201–202
Cerambyx cerdo	Great Capricorn beetle	–	Cerambycidae	Coleoptera	12, 37, 44, 77–78
Ernobius mollis	Bark waney edge borer	–	Ptinidae	Coleoptera	16, 17, 20, 148, 152, 171–175
Hexarthrum exiguum	Wood-boring weevil	Cossoninae	Curculionidae	Coleoptera	18, 20, 104, 107, 112, 113–117
Hylotrupes bajulus	House Longhorn beetle	–	Cerambycidae	Coleoptera	1, 12, 20, 21, 37, 50, 79–83, 189
Icosium tomentosum	Capricorn beetle	–	Cerambycidae	Coleoptera	14, 20, 37, 50, 73, 85–89
Ips typographus	Engraver beetle, European spruce bark beetle	Scolytinae	Curculionidae	Coleoptera	1, 2, 18, 20, 25, 124, 125, 126, 127–130
Kalotermes flavicollis	Yellow-necked dry-wood termite	–	Kalotermitidae	Isoptera	18, 19, 203, 205–208
Lyctus brunneus	Powder post beetle	–	Lyctidae	Coleoptera	3, 14, 17, 20, 139, 140, 142, 143–147, 203
Lyctus linearis	European Lyctus beetle	–	Lyctidae	Coleoptera	3, 14, 17, 20, 139, 140, 142, 143–147, 203
Nicobium castaneum	Library beetle	–	Ptinidae	Coleoptera	18, 19, 20, 148, 155, 177–181
Oligomerus ptilinoides	Furniture beetle woodworm	–	Ptinidae	Coleoptera	3, 18, 19, 20, 148, 156, 183–187
Penichroa fasciata	Capricorn beetle	–	Cerambycidae	Coleoptera	12, 20, 37, 56, 91–94
Pentarthrum huttoni	Wood-boring weevil	Cossoninae	Curculionidae	Coleoptera	18, 20, 104, 108, 112, 119–123
Scolytus multistriatus	Little elm bark beetle	Scolytinae	Curculionidae	Coleoptera	2, 18, 20, 124, 126, 131–134
Scolytus rugulosus	Fruit bark beetle	Scolytinae	Curculionidae	Coleoptera	2, 18, 20, 126, 135–138
Trogoxylon impressum	Powder post beetle	–	Lyctidae	Coleoptera	3, 14, 17, 20, 139, 141, 142, 143–147
Xestobium rufovillosum	Death watch beetle	–	Ptinidae	Coleoptera	16, 17, 20, 148, 157, 189–193
Xylocopa violacea	Large carpenter bee, Purple Xylocopa	–	Apidae	Hymenoptera	16, 17, 194, 195–198
Xylographus bostrichoides	Minute tree-fungus beetle	–	Ciidae	Coleoptera	18, 19, 20, 98, 99–103
Xylotrechus stebbingi	Capricorn beetle	–	Cerambycidae	Coleoptera	12, 20, 37, 69, 95–97

General References

Barbey A., 1923. La forêt européenne et sa résistance aux attaques des Insectes ravageurs. *Revue de botanique appliquée et d'agriculture coloniale*, 25, pp. 593-604.

Baslé K., 2011. Traitements de désinsectisation des biens patrimoniaux : législation et critères méthodologiques. *La lettre de l'OCIM*, 138, pp. 24-30.

Battisti A., 2004. Forests and climate change : lesson from insects. *Forest entomology*, the Italian Society of Siviculture and Forest Ecology, 1, pp. 17-24.

Beal J.-C., 1995. *L'arbre et la forêt, le bois dans l'antiquité*. De Boccard, publication de la bibliothèque Salomon-Reinach, Paris, 115 p.

Becker G., 1955. "Grundzüge der Insektensuccession in Pinus-Arten der Gebirge von Guatemala". *Journal of Applied Entomology*, 37 (1), pp. 1-28.

Becker G., 1971. "Physiological influences on wood-destroying insects of wood compounds and substances produced by microorganisms". *Wood Science and Technology*, 5, pp. 236–246. https://doi.org/10.1007/BF00353686

Bell CH., 2014. "Infestation Management in Food Production Premises". Food safety Assurance systems. *Encyclopedia of Food Safety*, 4, pp. 189-200.

Berti S., Crivellaro A., Nocetti M., Rescic L., Sozzi L., 2007. *Conoscere il legno*. CNR Ivalsa, Via del legno, CNR e Consorzio Boschi Carnici, 21 p.

Bertrame I. N., 2017. L'insecte à l'œuvre. *Techniques et cultures*, 68 (2), pp. 162-177.

Bevan D., 1987. *Forest insects*. HMSO Books, 153 p.

Blanchette R. A., 1991. "Deterioration in historic and archaeological woods from terrestrial sites." *In:* Koestler R. J., Koestler V. H., Charola A. E., Nieto-Fernandez F. E. (eds). *Art, biology, and conservation: biodeterioration of works of art.* En anglais normalement c'est Metropolitan Museum of Art, New-York, pp. 328-347.

Bobadilla L., Arriaga F., Luengo E., Martinez R., 2015. "Dimensional and morphological analysis of the detritus from six European wood boring insects". *Maderas, Ciencia y tecnologia*, 17 (4), pp. 893-904.

Botton B., Breton A., Fevre M., Gauthier S., Guy P., Larpent J.-P., Reymond P., Sanglier J.-J., Vayssier Y., Veau P., 1990. *Moisissures utiles et nuisibles : importance industrielle.* Masson, 2e édition, collection Biotechnologies, Paris, 512 p.

Bouget C., Brustel H., Nageleisen L. M., 2005. Nomenclature des groupes écologiques d'insectes liés au bois : synthèse et mise au point sémantique. *C. R. Biologies*, 328, pp. 936-948.

Brazier J., Franklin G. L., 1961. "Identification of Hardwoods: a microscope key", *Forest Products Research*, n°46, London, Department of scientific and industrial Research, 96 p.

Bremond J., Lessertisseur J., 1973. Lamarck et l'entomologie. *Revue d'histoire des sciences*, 26 (3), pp. 231-250.

Buckland P. C., Coope G. R., 1991. *A bibliography and literature review of Quaternary Entomology.* Sheffield: J. R. Collis Publications, Dept. of Archaeology & Prehistory, University of Sheffield, 24 p.

Buse J., Levanony T., Timm A., Dayan T., Assmann T., 2010. "Saproxylic beetle assemblages in the Mediterranean region: Impact of forest management on richness and structure". *Forest Ecology and Management*, 259, pp. 1376-1384.

Cambefort Y., 2006. *Des coléoptères, des collections & des hommes.* Publications Scientifiques du Muséum national d'Histoire naturelle, Paris, 375 p.

Campbell G., Kenward H., 2012. "Insect and plant remains". *In:* Tipper J. (ed) *Experimental Archaeology and Fire: the investigation of a burnt reconstruction at West Stow Anglo-Saxon Village,* East Anglian Archaeology, pp 90-181.

Centre Technique du Bois et de l'Ameublement, 1996. *Le traitement curatif des bois dans la construction.* Département BIOTEC (Biologie, Environnement, Technologies). Eyrolles, Paris, 140 p.

Couvert M., 1977. *Atlas d'anatomie des charbons de foyers préhistoriques. Afrique du Nord tempérée, Mémoires du centre de recherches anthropologiques préhistorique et ethnographiques.* Alger, SNED, Mémoires du CRAPE, vol. 26, 28 p.

Curling S., 2017. "Test methods for bio-based building materials". *In:* Jones D., Brischke C., Performance of Bio-based Building Materials, *Woodhead Publishing*, pp. 385-481. https://doi.org/10.1016/B978-0-08-100982-6.00007-0

Dajoz R., 1980. *Écologie des insectes forestiers.* Gauthier-Villars, Bordas, Paris, 489 p.

Dajoz R., 2007. *Les insectes et la forêt.* TEC & DOC, 2e édition, Rôle et diversité des insectes dans le milieu forestier, Lavoisier, Paris, 648 p.

Da Lage A., Métailié G., 2005. *Dictionnaire de biogéographie végétale*, CNRS éditions, Paris, 580 p.

Dirol D., Deglise X., 2001. *Durabilité des bois. Mécanique et ingénierie des matériaux.* Hermès, science publications, Paris, 416 p.

Dodelin B., 2006. *Écologie et biocénoses des coléoptères saproxyliques dans quatre forêts du nord des Alpes françaises.* PhD, Université de Savoie, Chambéry, Annecy, 159 p.

Dop P., Gauthier A., 1928. *Manuel de technique botanique, histologie et microbiologie végétales.* J. Lamarre, pp. 101-115.

Elvira M., Hernando L., 1989. *Inflamabilidad y energía de las especies de sotobosque.* Instituto Nacional de Investigaciones Agrarias, Ministerio de Agricultura, Pesca y Alimentación: Madrid, INIA Monographs, 99 p.

Findlay W. P. K., 1967. *Timber Pests and Diseases: Pergamon Series of Monographs on Furniture and Timber.* Pergamon, Elsevier, 292 p.

Fischesser B., 1996. *Connaître les arbres.* Nathan Nature, Techniguides, Paris, 351 p.

Florian M. L., 1997. *Heritage Eaters : insects and fungi in heritage collection.* James&James (Science Publishers), London, 164 p.

Fohrer F., 2011. Le diagnostic des infestations en milieu patrimonial : approches techniques et méthodologiques. *La lettre de l'OCIM*, 38, pp. 31-40

Fraval A., 2008. Les coléoptères du bois ouvré. *Insectes*, 151 (4), pp. 29-33.

Gambetta A., 2010. *Funghi e insetti nel legno. Diagnosi, prevenzione, controllo (Italian) Paperback.* Nardini Editore, Libro universitario, 155 p.

Gaonkar C. A., Chandrashekar Anil A., 2013. Stable isotopic analysis of Barnacle larvae and their faecal pellets to evaluate the ingested food *Journal of Experimental Marine Biology and Ecology,* 441, pp. 28-32.

Gärtner H., Schweingruber F. H., 2013. *Microscopic Preparation Techniques for Plant Stem Analysis.* Design & layout by Börner A., Birmensdorf 78 p.

Gärtner H., Lucchinetti S., Schweingruber F. H., 2014. "New perspectives for wood anatomical analysis in dendrosciences : the GSL1 – microtome". *Dendrochronologia*, 32, pp. 47-51.

Giustina G. D., 1985. *La pathologie des charpentes en bois, causes des désordres et remèdes.* Ed. du Moniteur, 143 p.

Greguss P., 1955. *Identification of living Gymnosperms on the basis of xylotomy.* Akademiai Kiado, Budapest, 263 p.

Hedges S. B., 2012. "Wormholes record species history in space and time". *Biology letters*, 9, Pennsylvania, U.S.A., 4 p. https://doi.org/10.1098/rsbl.2012.0926

Hickin N. E., 1963. *The Insect Factor in Wood Decay.* Hutchinson and Co., London, 336 p.

Hickin N. E., 1967. *The conservation of building timbers.* The Rentokil library, Hutchinson of London, London, 140 p.

Huber B., Rouschal C., 1954. *Mikrophotographischer Atlas mediterraner Hölzer.* Berlin-Grunewald, Fritz Haller Verlag, 105 p.

Jacquiot C., Trenard Y., Dirol D., Boureau M. E., 1973. *Atlas d'anatomie des bois des Angiospermes.* Centre technique du bois, Paris, vol I, 75 p.

Jullien J., 2008. Insectes xylophages en pépinière de jeune plant. *PHM, Revue Horticole, la revue technique des pépiniéristes, horticulteurs et maraîchers*, 508, pp. 14-20.

Kenward H., Carrott J., 2006. "Insect species associations characterise past occupation sites". *Journal of Archaeological Science*, 33, pp. 1452-1473.

Kenward H., 2006. "The visibility of past trees and woodland: testing the value of insect remains". *Journal of Archaeological Science*, 33, pp. 1368-1380.

Kingsley H., Pinniger D., Xavier-Rowe A., Winsor P., 2001. *Integrated pest management for collections.* Proceedings of 2001: a pest Odyssey English Heritage, 150 p.

Kollmann F., Côté W. A., 1968. *Principes of Wood Science and Technology.* vol I. *Solid wood*, 1. Springer-verlag, New York, 592 p.

Kurtz O'Deal L., Harris K. L., 1962. *Micro-analytical Entomology for Food Sanitation Control.* Association of Official Agricultural Chemists, Washington, 576 p.

Lavisci P., 2001. *Pathologies des structures en bois : analyse des caractéristiques physiques et mécaniques des collages structuraux pour la restauration des charpentes.* PhD Science du bois (dir) P. Triboulot, Nancy 1, 171 p.

Le Conte S., Vaiedelich S., Thomas J. H., Mulavia V., De Reyer D., Maurin E., 2015. "Acoustic emission to detect xylophagous insects in wooden musical instrument". *Journal of Cultural Heritage,* vol. 16, pp. 338-343.

Lepesme P., 1944. *Les Coléoptères des denrées alimentaires et des produits industriels entreposés.* Paul Lechevalier, Paris, 124 p., 233 fig., 12 pl.

Martinez M., 2008. Les insectes xylophages : Qui sont-ils ? Que mangent-ils ? *PHM, Revue Horticole, la revue technique des pépiniéristes, horticulteurs et maraîchers*, 508, pp. 11-13.

Meusnier I., Martinez M., Fohrer F., 2017. Utilisation des outils moléculaires pour l'identification des insectes nuisibles au patrimoine : du rêve à la réalité. Monographie du colloque Croqueurs de patrimoine : les infestations entomologiques : enjeux d'aujourd'hui, politique de demain. *Les dossiers de l'Ocim*, pp. 195-208.

Micas L., 2011. Commentaires sur les Coléoptères saproxyliques découverts lors de l'inventaire de la réserve biologique du Luberon (Vaucluse). Deuxième partie : Coleoptera autres que Cerambycidae. *L'Entomologiste*, 67 (5), pp. 257-260.

Moret P., 1998. L'archéo-entomologie, ou les insectes au service de l'histoire. *Revue de médecine vétérinaire*, 149 (11), pp. 995-998.

Moskal-del Hoyo M., Wachowiak M., Blanchette R. A., 2010. "Preservation of fungi in archaeological charcoal". *Journal of Archaeological Science*, 37, pp. 2106-2116.

Nageleisen L. M., Bouget C., 2009. L'étude des insectes en forêt : méthodes et techniques, éléments essentiels pour une standardisation : synthèse des réflexions menées par le groupe de travail "Inventaires Entomologiques en Forêt". *Les Dossiers Forestiers*, 19, Office National des Forêts, Fontainebleau, 144 p.

Nageleisen L. M., Saintonge F. X., Piou D., Riou-Nivert P., 2010. *La santé des forêts. Maladies, insectes, accidents climatiques. Diagnostic et prévention.* Département de la santé des forêts. Institut pour le développement forestier, Paris, 608 p.

Nardi G., Ratti E., 1995. "*Coleoptera Lyctidae, Bostrychidae e Anobiidae di Pantelleria e Lampedusa*". *In:* Massa B., (Ed.) *Arthropoda di Lampedusa, Linosa e Pantelleria (Canale di Sicilia, Mar Mediterraneo).* Il Naturalista siciliano 19 (Supplemento), pp. 477-492.

New T. R., 2016. *Alien Species and Insect Conservation.* Springer International Publishing Switzerland, 239 p. https://doi.org/10.1007/978-3-319-38774-1

Nobre T., Rouland-Lefèvre C., Aanen D. K., 2011. *Comparative Biology of Fungus Cultivation in Termites and Ants. In:* Bignell D., Roisin Y. et Lo N. (eds), *Biology of Termites a Modern Synthesis.* Springer, pp. 193-210.

Paine T. D., Lieutier F., 2016. *Insects and Diseases of Mediterranean Forest Systems.* Springer International Publishing Switzerland, 892 p. https://doi.org/10.1007/978-3-319-24744-1

Palma P., Santhakumaran L. N., 2014. *Shipwrecks and global worming.* Archaeopress, Oxford, 65 p.

Paulian R., 1988. *Biologie des coléoptères.* Centre National des lettres. Editions Lechevalier, Paris, 719 p.

Pinniger D., 2001. *Pest Management in Museums, Archives and Historic Houses.* Ed. Archetype Publication, London, 115 p.

Pitman A. J., Jones A. M., Gareth Jones E. B., 1993. "The wharf-borer *Nacerdes Melanura* L., a threat to stored archaeological timbers". *Studies in conservation*, 38, pp. 274-284.

Ponel P., Yvinec J.-H., 1997. L'archéomentomologie en France. *Les nouvelles de l'archéologie*, 68, pp. 31-37.

Portevin G., 1929. *Histoire Naturelle des coléoptères de France. Tome I: Adephaga-Polyphaga: staphylinoidea.* Encyclopédie entomologique, série A, XII, Ed. Paul Lechevalier & Fils, Paris, 649 p.

Portevin G., 1931. *Histoire Naturelle des coléoptères de France. Tome II: Polyphaga: Lamellicornia, palpicornia, diversicornia.* Encyclopédie entomologique, série A, XIII, Ed. Paul Lechevalier & Fils, Paris, 542 p.

Portevin G., 1934. *Histoire Naturelle des coléoptères de France. Tome III: Polyphaga: Heteromera, phytophaga.* Encyclopédie entomologique, série A, XVII, Ed. Paul Lechevalier & Fils, Paris, 374 p.

Portevin G., 1935. *Histoire Naturelle des coléoptères de France. Tome IV: Polyphaga: Rhynchophora.* Encyclopédie entomologique, série A, XVIII, Ed. Paul Lechevalier & Fils, Paris, 500 p.

Randall C. J., 2000. *Management of Wood-destroying Pests. A Guide for Commercial Applicators Category 7B.* Extension Bulletin E-2047, Major revision-destroy old stock, Michigan State University Extension, 124 p.

Riesco Muñoz G., Remacha Gete A., Gasalla Regueiro M., 2014. "Sawing yield in oak (*Quercus robur*) wood affected by insect damage". *International Biodeterioration & Biodegradation*, 86, pp. 102-107.

Ruffinatto F., Crivellaro A., 2019. *Atlas of macroscopic wood identification with a special focus on timbers used in Europe and CITES-listed species.* Springer Nature Switzerland AG, Cham, Switzerland, 439 p. https://doi.org/10.1007/978-3-030-23566-6

Samson R. A., Hoekstra E. S., Frisvad J. C., Filtenborg O., 2002. *Introduction to food-and airborne fungi.* 6e édition, Centraalbureau voor schimmelcultures, An institute of the Royal Netherlands Academy of Arts and Sciences, Utrecht, 389 p.

Schiegg Pasinelli K., Suter W., 2002. *Lebensraum Totholz.* 2. Aufl. Merkbl. Prax. 6 p.

Schweingruber F. H., 1990. *Anatomie europäischer Hölzer. Anatomy of European woods.* WSL-FNP, Haupt, 800 p.

Seibold S., Müller J., Baldrian P., Cadotte M. W., Štursová M., Krah F.-S., Bässler C., 2018. "Fungi associated with beetles dispersing from dead wood - Let's take the beetle bus!" *Fungal Ecology*, 39, pp. 100-108.

Serment M. M., 1981. *The most important wood-destroying insects in various countries (results of questionnaire).* Centre technique du bois, Paris, The international Research group of wood preservation, biological problems, document n° IRG/WP/1136.

Stušek P., Pohleven F., Dušan C., 2000a. "Detection of wood boring insects by measurement of oxygen consumption". *International Biodeterioration & Biodegradation*, 46, pp. 293-298.

Therond J., 1975a. Catalogue des coléoptères de la Camargue et du Gard. Première partie : Caraboidea, Palpicornia, Staphylinoidea, Diversicornia, Heteromera, Lamellicornia. Société d'étude des Sciences Naturelles de Nîmes. *Mémoire* (10), 411 p.

Therond J., 1975b. Catalogue des coléoptères de la Camargue et du Gard. Deuxième partie : Phytophaga : Cerambycidae, Chrysomelidae, Bruchidae, Rhynchophora : Anthribidae, Brenthidae, Curculionidae, Scolytidae. Société d'étude des Sciences Naturelles de Nîmes. *Mémoire* (10), 193 p.

Toriti M., 2018. *Les bois ouvragés en Gaule romaine : approches croisées archéologiques, anthraco-xylologiques et entomologiques*, PhD Thesis, Le Mans University, unpublished, 1052 p. https://halshs.archives-ouvertes.fr/tel-02382217

Vernet J.-L., coll. Ogereau P., Figueral I., Machado Zanes C., Uzquiano P., 2001. *Guide d'identification des charbons de bois préhistoriques et récents, Sud-Ouest de l'Europe: France, péninsule ibérique, et îles Canaries.* CNRS Editions, Paris, 395 p.

Weiss M. R. 2006. "Defecation Behavior and Ecology of Insects". *Annu. Rev. Entomol.* 51, pp. 635-61. https://doi.org/10.1146/annurev.ento.49.061802.123212

Winsor P., Pinniger D., Bacon L., Child B., Harris K., Lauder D., Phippard J., Xavier-Rowe A., 2011. *Integrated pest management for collections.* Proceedings of 2011 : a pest odyssey, 10 years later. English Heritage, 223 p.

Yela J.-L., 1997. "Insectos causantes de daños al patrimonio historico y cultural : caracterizacion, tipos de daño y methodos de lucha (Arthropoda : Insectos)". *Los Artropodos y el Hombre, Bol. S.E.A.*, 20, pp. 111-122.

Specialized References

Apidae

Banaszak J., Banaszak-Cibicka W., Twerd L., 2019. "Possible expansion of the range of *Xylocopa violacea* L. (Hymenoptera, Apiformes, Apidae) in Europe". *Turkish Journal of Zoology*, 43, pp. 650-656. https://doi.org/10.3906/zoo-1812-6

Dindo M. L., Campadelli G., Gambetta A., 1991. "Note su *Xylocopa violacea* L. e *Xylocopa valga* Gerst. (Hym. Anthophoridae) nidificanti nei tronchi della foresta fossile di Dunarobba (Umbria)". *Boll. Ist. Ent. G. Grandi Università di Bologna*, 46, pp. 153-160.

Gerling D., Velthuis H.H.W., Hefetz A., 1989. "Bionomics of the large carpenter bees of the genus *Xylocopa*". *Annu. Rev. Entomol.*, 34, pp. 163-190.

Gonzalez V. H., Hrnitz J. M., Percival C. R., Pulley K. L., Tapsak S. T., Tscheulin T., Petanidou T., Barthell J. F., 2020. "Thermal tolerance varies with dim-light foraging and elevation in large carpenter bees (Hymenoptera: Apidae: Xylocopini)". *Ecological Entomology*, 45, pp. 688-696.

Vicidomini S., 1996. "Biology of *Xylocopa violacea* (Hymenoptera): In nest ethology". *Italian Journal of Zoology*, 63 (3), pp. 237-242. https://doi.org/10.1080/11250009609356139

Bostrichidae

Baena M., Zuzarte A. J., 2013. "Contribution to the study of Bostrichidae from Portugal and update of the catalogue of the Iberian fauna (Coleoptera, Bostrichidae)". *Zool. baetica*, 24, pp. 25-51.

Bieńkowski A. O., Orlova-Bienkowskaja M. J., 2017. "Establishment of the invasive pest of bamboo *Dinoderus japonicus* Lesné, 1895 (Coleoptera Bostrichidae) in the Caucasus and notes on other beetle species currently established in this region". *Redia*, 100, pp. 115-118. https://doi.org/10.19263/REDIA-100.17.14

Bonsignore C. P., 2012. *"Apate monachus* (Fabricius, 1775), a Bostrichid Pest of Pomegranate and Carob Trees in Nurseries" Short Communication. *Plant Protect. Sci.*, 48 (2), pp. 94–97.

Borowski J., 2007a. *"Bostrichidae"*. *In:* Löbl I., Smetana A., (eds). *Catalogue of Palaearctic Coleoptera. Volume 4. Elateroidea – Derodontoidea – Bostrichoidea – Lymexyloidea – Cleroidea – Cucujoidea.* Apollo Books, Stenstrup, pp. 320-328.

Borowski J., 2013. *"Errata for volume 4 [Bostrichidae]"*. *In:* Lobl I., Smetana A., (eds). *Catalogue of Palaearctic Coleoptera. Volume 8. Curculionoidea II.* Apollo Books, Stenstrup, pp. 33-34.

Borowski J., 2018. "Materials to the knowledge of Bostrichidae (Coleoptera) of The Republic of Gambia". *World Scientific News*, 106, pp. 1-11.

Borowski J., Gutowski J. M., Slawski M., Sućko K., Zub K., 2018. *"Stephanopachys linearis* (Kugelann, 1792) (Coleoptera, Bostrichidae) in Poland". *Nature Conservation*, 27, pp. 75–84. https://doi.org/10.3897/natureconservation.27.24977

Brustel H., Gouix N., Bouyon H., Rogé J., 2013. Les *Stephanopachys* de la faune ouest-paléarctique (Coleoptera Bostrichidae) : distribution et reconnaissance des trois espèces françaises au service de l'application de la directive Habitats, Faune, Flore. *L'Entomologiste*, 69 (1), pp. 41-50.

Buchelos C. T., 1991. *"Dinoderus minutus* and *D. brevis* (Coleoptera: Bostrychidae): Two Exotic Wood Borers Introduced to Greece". *Entomologia Hellenica*, 9, pp. 13-15. https://doi.org/10.12681/eh.13987

Chararas C., Balachowsky R., 1962. *Famille des Bostrychidae. In:* Balachowsky R. (Ed.) *Entomologie appliquée à l'agriculture. Tome I. Coléoptères premier volume.* Mason et Cie., Paris, pp. 304-315.

Cymorek S., 1979. "Data sheet on woodboring insects : *Bostrichus capucinus* (Linneus)". The international Research Group on Wood Preservation. *Biological problems,* IRG/WP/193, Scotland, 14 p.

Lachat T., Brang P., Bolliger M., Bollmann K., Brändli U. B., Bütler R., Herrmann S., Schneider O., Wermelinger B., 2019. "Totholz im Wald". Entstehung, Bedeutung und Förderung. 2. überarbeitete *Aufl. Merkbl. Prax.*, 52, 12 p.

Liu L.-Y., Ghahari H., Beaver R. A., 2016. "An annotated synopsis of the powder post beetles of Iran (Coleoptera: Bostrichoidea: Bostrichidae)". *Journal of insect Biodiversity*, 4 (14), pp. 1-22.

Liu L.-Y., Geis K.-U., 2019. "A synopsis of the Lyctine beetles of Eurasia with a key to the species (Insecta: Coleoptera: Bostrichidae: *Lyctinae)"*. *Journal of Insect Biodiversity*, 9 (2), pp. 34-56. https://doi.org/10.12976/jib/2019.09.2.1

López-Pérez J.-J., 2012. "Los Bostrichidae Latreille, 1802 (Coleoptera, Bostrichoidea) de la provincia de Huelva (S. O. de Andalucía, España)". *Revista gaditana de Entomología*, 3(1-2), pp. 23-28.

Luna Murillo A., Obregón R., 2013. "Nuevas aportaciones a la fauna de Bostrichidae (Coleoptera) de la provincia de Córdoba (Andalucía, España)". *Boletín de la SAE*, 21, pp. 46-57.

Marzo L. de, Porcelli F., 1987. "Struttura delle «aree sensoriali» antennali di alcuni Bostrichidi (Coleoptera)". *Entomologica*, XXII, Bari, 20-XII, pp. 87-95.

Muscarella C., Sparacio I., Liberto A., Nardi G., 2013. "The genus *Lichenophanes* Lesne, 1899 in Italy (Coleoptera Bostrichidae) and short considerations on the saproxylophagous beetle-fauna of Nebrodi Mountains (Sicily)". *Biodiversity Journal*, 4 (4), pp. 451-466.

Nardi G., Badano D., De Cinti B., 2015. "First record of *Dinoderus* (*Dinoderastes*) *japonicus* in Italy (Coleoptera: Bostrichidae)". *Fragmenta entomologica*, 47 (2), pp. 147-150.

Nardi G., Mifsud D., 2015. "The Bostrichidae of the Maltese Islands (Coleoptera)". *ZooKeys*, 481, pp. 69–108. https://doi.org/10.3897/zookeys.481.8294

Nardi G., Audisio P., 2016. "Italian account for *Stephanopachys linearis* (Kugelann, 1792), a species listed in Annex II of the Habitats Directive (Coleoptera: Bostrichidae)". *Fragmenta entomologica*, 48 (2), pp. 131-136.

Ozgen I., 2018. "New findings" on *Bostrichus capucinus* Linnaeus, 1758 (Coleoptera: Bostrichidae) in Turkey". *Fresenius Environmental Bulletin*, 25 (5), pp. 2829-2833.

Parmar A., Kirchner S. M., Langguth H., Döring T. F., Hensel O., 2017. "Boxwood Borer *Heterobostrychus brunneus* (Coleoptera: Bostrichidae) Infesting Dried Cassava: A Current Record from Southern Ethiopia". *Journal of Insect Science*, 17 (1): 14, pp. 1-8. https://doi.org/10.1093/jisesa/iew106

Sangwook P., Seunghwan L., Ki-Jeong H., 2015. "Review of the family Bostrichidae (Coleoptera) of Korea". *Journal of Asia-Pacific Biodiversity*, 8, pp. 298-304. https://doi.org/10.1016/j.japb.2015.10.015

Savoldelli S., Regalin R., 2009. "Infestation of wood pallers by *Sinoxylon unidentatum* (Fabricius) (Coleoptera Bostrichidae) in Italy". *Boll. Zool. agr. Bachic. Ser.*, II, 41 (3), pp. 235-238.

Cerambycidae

Allemand R., Brustel H., Clary J., 2002. Une espèce de Cerambycidae nouvelle pour la faune de France, *Aegomorphus francottei* Sama (Coleoptera). *Bulletin mensuel de la Société linnéenne de Lyon,* 71e année, 3, pp. 105-114. https://doi.org/10.3406/linly.2002.13376

Allison J. D., Borden J. H., Seybold S. J., 2004. "A review of the chemical ecology of the Cerambycidae (Coleoptera)". *Chemoecology*, 14, pp. 123-150. https://doi.org/10.1007/s00049-004-0277-1

Anichtchenko A., Verdugo Páez A., 2004. "Iberodorcadion (Hispanodorcadion) zenete, nueva especie ibérica de cerambícido (Coleoptera, Cerambycidae) procedente de Sierra Nevada (Andalucía, España)". *Boletín de la Sociedad Andaluza de Entomología*, 11, pp. 31-42.

Berger P., 2012. *Coléoptères Cerambycidae de la faune de France continentale et de Corse.* Actualisation de l'ouvrage d'André Villiers, 1978. Ed. Association Roussillonaise d'entomologie (A.R.E.), Perpignan, 664 p.

Bétard F., Gerbaud J., 2013. Sur quelques coléoptères Cerambycidae remarquables observés au Rocher de Cheffois (Vendée, France). *Le naturaliste vendéen*, 11, pp. 69-75.

Bily S., Mehl O., 1989. *Longhorn Beetles - Coleoptera, Cerambycidae - Of Fennoscandia and Denmark.* Brill, Fauna Entomologica Scandinavica Series, 208 p.

Brustel H., Van Meer C., 2001. Données originales sur quelques Cerambycidae des Pyrénées et régions voisines (Coleoptera). *Bulletin de la Société linnéenne*, Bordeaux, 29 (1), pp. 11-20.

Brustel H., Berger P., Cocquempot C., 2002. Catalogue des Vesperidae et des Cerambycidae de la faune de France. *Annales de la Société Entomologique de France*, 38 (4), pp. 443-461. https://doi.org/10.1080/00379271.2002.10697354

Busconi M., Berzolla A., Chiappini E., 2014. "Preliminary data on cellulase encoding genes in the xylophagous beetle, Hylotrupes bajulus (Linnaeus)". *International Biodeterioration & Biodegradation*, 86, pp. 92-95.

Calmont B., Sautière C., 2007. Première citation de *Penichroa fasciata* (Stephens, 1831) dans le département de l'Ardèche (Coleoptera Cerambycidae). *L'Entomologiste*, 63 (2), pp. 55-56.

Cocquempot C., Soldati F., Parmain G., 2012. *Xylotrechus stebbingi* (Gahan, 1906) nouveau pour le département de l'Aude (Coleoptera, Cerambycidae). *R.A.R.E.*, T. XXI (2), pp. 1-2.

Cocquempot C., Desbles F., Mouttet R., Valladares L., 2019. *Xylotrechus chinensis* (Chevrolat, 1852), nouvelle espèce invasive pour la France métropolitaine (Coleoptera, Cerambycidae, *Clytini*). *Annales de la Société Entomologique de France*, 124 (1), pp. 27-32.

Crivellaro A., 2005. "Note su *Trichoferus holosericeus* (Rossi, 1790) (Coleoptera cerambycidae), xilofago del legno secco". *Redia*, LXXXVIII, pp. 109-113.

Doychev D., Georgiev G. T., 2006. "Notes on distribution and ecology of *Icosium tomentosum* atticum Ganglbauer (Coleoptera: Cerambycidae) in Bulgaria". *Forest Science*, 3, pp. 117-122.

Drumont A., Smets K., Scheers K., Thomaes A., Vandenhoudt R., Lodewyckx M., 2014. *Callidiellum rufipenne* (Motschulsky, 1861) en Belgique : bilan de sa présence et de son installation sur notre territoire (Coleoptera : Cerambycidae : Cerambycinae). *Bulletin de la Société royale belge d'Entomologie/Bulletin van de Koninklijke Belgische Vereniging voor Entomologie*, 150, pp. 239-249.

Duelli P., Wermelinger B., 2010. "Der Alpenbock (*Rosalia alpina*)". Ein seltener Bockkäfer als Flaggschiff-Art. 2. *überarbeitete Aufl. Merkbl. Prax.*, 39, 8 p.

Gambetta A., Orlandi E., 1988. "*Penichroa fasciata* (Stephens) (Col. Cerambycidae) a pest in wood materials". *The international research group on wood preservation*, IRG/WP., 1365, 3 p.

Grancher C., 2013. Note on some interesting saproxylic Coleoptera found on the commune of Sault-de-Navailles (Pyrénées-Atlantiques). *Bull. Soc. Linn.*, Bordeaux, 148, nouvelle série 41 (1), pp. 7-12.

Guérard P., 2002. Les longicornes de la Manche catalogue illustré et cartographie. *L'Argiope*, 38, pp. 27-53.

Hanks L. M., 1999. "Influence of the larval host plant on reproductive strategies of cerambycid beetles". *Annu. Rev. Entomol.*, 44, pp. 483-505.

Hellrigl K., 2006. "Über Auftreten von Holzinsekten in Häusern". *Forest observer* (2-3), pp. 333-348.

Lazarev M. A., 2009. "*Cornumutila quadrivittata* (Gebler, 1830) and *C. lineata* (Letzner, 1844), stat. rest. (Coleoptera, Cerambycidae) from Western Europe and Russia". *Spec. Bull. Jpn. Soc. Coleopterol.*, Tokyo, 7, pp. 117-126.

Linsley E. G., 1959. "Ecology of Cerambycidae". *Annu. Rev. Entomol.*, 4, pp. 99-138.

Martikainen P., 2002. "Ecology and conservation status of *Acanthocinus griseus* (Fabricius, 1792) (Coleoptera: Cerambycidae) in Finland". *Entomol. Fennica*, 13, pp. 41–50.

Meshkova V., Zinchenko O., 2017. "Seasonal development of the timberman beetle *Acanthocinus aedilis* (Linnaeus, 1758) (Coleoptera: Cerambycidae) in the North-Eastern Steppe of Ukraine". *The Kharkov Entomological Society Gazette*, 25 (2), pp. 40-44. ISSN 1726–8028

Ostojá-Starzewski J. C., 2016. "Red-Necked Longhorn Beetle -*Aromia bungii*". Plant Pest Factsheet, 6 p.

Rossa R., Goczal J., 2017. "Hind wing variation in *Leptura annularis* complex among European and Asiatic populations (Coleoptera, Cerambycidae)". *ZooKeys*, 724, pp. 31–42. https://doi.org/10.3897/zookeys.724.20667

Stejskal V., Douda O., Zouhar M., Manasova M., Dlouhy M., Simbera J., Aulicky R., 2014. "Wood penetration ability of hydrogen cyanide and its efficacy for fumigation of Anoplophora glabripennis, *Hylotrupes bajulus* (Coleoptera), and *Bursaphelenchus xylophilus* (Nematoda)". *International Biodeterioration & Biodegradation,* 86, pp. 189-195.

Stušek P., Pohleven F., Dušan C., 2000b. "Detection of wood boring insects by measurement of oxygen consumption". *International Biodeterioration & Biodegradation*, 46, pp. 293-298.

Tăuşan I., Bucşa C., 2010. "Palaearctic longhorn beetles (Coleoptera: Cerambycidae) from "Dr. Karl Petri" collection of the Natural History Museum of Sibiu (Romania). Part I: Lepturinae subfamily". *Travaux du Muséum National d'Histoire Naturelle Grigore Antipa*, 53, pp. 223-233. https://doi.org/10.2478/v10191-010-0017-4

Touroult J., Cima V., Bouyon H., Hanot C., Horellou A., Brustel H., 2019. *Longicornes de France – Atlas préliminaire (Coleoptera : Cerambycidae & Vesperidae).* Supplément au bulletin d'ACOREP-France, Paris, 176 p.

Van Meer C., Cocquempot C., 2013. Découverte d'un foyer de *Callidiellum rufipenne* (Motschulsky, 1861) dans les Pyrénées-Atlantiques (France) et correction nomenclaturale (Cerambycidae Cerambycinae Callidiini). *L'Entomologiste*, 69 (2), pp. 87-95.

Vitali F., 2004. "*Xylotrechus smei* (Castelnau & Gory, 1841): its presence in Western Palaearctic region and description of the pupa (Coleoptera, Cerambycidae)". *Supplemento agli Annali del Museo civico di storia naturale G Doria*, Genova, Res Ligusticae CCXLIII, 7 (340), pp. 1-8. ISSN 0417 - 9927

Wermelinger B., Forster B., Hölling D., Plüss T., Raemy O., Klay A., 2015. "Invasive Laubholz-Bockkäfer aus Asien". *Ökologie und Management.* 2. überarbeitete Aufl. Merkbl. Prax. 50, 16 p.

Ciidae

Benick L., 1952. "Pilzkäfer und Käferpilze, Ökoligische und Statistische Untersuchungen". *Acta Zoologica Fennica*, 70, pp. 1-250.

Birkemoe T., Jacobsen R. M., Sverdrup-Thygeson A., Biedermann P. H. W., 2018. *"Insect-Fungus Interactions in Dead Wood Systems". In:* Ulyshen M. (eds) *Saproxylic Insects.* Zoological Monographs, vol 1. Springer, Cham, pp. 377-427. https://doi.org/10.1007/978-3-319-75937-1_12

Bouget C., Brustel H., Nageleisen L. M., 2005b. Nomenclature des groupes écologiques d'insectes liés au bois : synthèse et mise au point séman-tique. *Comptes-Rendus Biologies*, 328 (10-11), pp. 936-948.

Fäldt J., Jonsell M., Norlander G., Borg-Karlson A. K., 1999. "Volatiles of bracket fungi *Fomitopsis pinicola* and *Fomes fomentarius* and their functions as insect attractants". *Journal of Chemical Ecology*, 25 (3), pp. 567-590.

Fossli T. E., Andersen J., 1998. "Host preference of Cisidae (Coleoptera) on tree-inhabiting fungi in northern Norway". *Entomologica Fennica*, 9 (2), pp. 65-78.

Gielen K., 2018. "*Cis bilamellatus* (Wood, 1884) and *Xylographus bostrichoides* (Dufour, 1843), two new minute tree-fungus beetles for the Belgian fauna (Coleoptera: Ciidae)". *Bulletin de la Société royale belge d'Entomologie/Bulletin van de Koninklijke Belgische Vereniging voor Entomologie*, 154, pp. 242-246.

Graves R. C., 1960. "Ecological observations on the insects and other inhabitants of woody shelf fungi (Basidiomycetes: Polyporaceae) in the Chicago area". *Annals of the Entomological Society of America*, 53, pp. 61-78.

Jonsell M., Norlander G., 1995. "Field attraction of Coleoptera to odours of the wood-decaying polypores *Fomitopsis pinicola* and *Fomes fomentarius*". *Annales Zoologici Fennici*, 32 (4), pp. 391-402.

Komonen A., Kouki J., 2005. "Occurrence and abundance of fungus-dwelling beetles (Ciidae) in boreal forest and clearcuts: habitat associations at two spatial scales". *Animal Biodiversity and Conservation*, 28 (2), pp. 137-147.

Lawrence J. F., 1973. "Host reference in ciidbettles (Coleoptera: Ciidae) inhabiting the fruiting bodies of Basidiomycetes in North America". *Bulletin of the Museum of Comparative Zoology*, 145 (3), pp. 163-212.

Orledge G. M., Reynolds S. E., 2005. "Fungivore host-use groups from cluster analysis: patterns of utilisation of fungal fruiting bodies by ciid beetles". *Ecological Entomology*, 30 (6), pp. 620-641.

Rose O., 2012. Les Ciidae de la faune de France continentale et de Corse : mise à jour de la clé des genres et du catalogue des espèces (Coleoptera Tenebrionoidea). *Bulletin de la Société entomologique de France*, 117 (3), pp. 339-362.

Thakeow P., Angeli S., Weissbecker B., Schütz S., 2008. "Antennal and behavioral responses of Cisboleti to fungal odor of *Trametes gibbosa*". *Chemical Senses*, 33 (4), pp. 379-387.

Curculionidae

Chararas C., 1957. *Étude anatomique et biologique de quelques Curculionidae xylophages et comparaison avec des Scolytidae*. PhD, Paris, N° d'ordre 523, 75 p.

Conord C., 2006. *Ecologie, génétique et symbiose bactérienne chez le grand charançon du pin*, Hylobius abietis : *adaptation d'un insecte ravageur à son environnement forestier*. Ecologie, Environnement. Université Joseph-Fourier - Grenoble I, PhD, 196 p.

Crooke M., 1947. "*Pissodes Validirostris* Gyll., in shoots of *Pinus sylvestris* L." *Forestry: An International Journal of Forest Research*, 21 (2), pp. 221–226. https://doi.org/10.1093/oxfordjournals.forestry.a062882

Garcia R., Andukar C., Oromi P., Emerson B., Lopez H., 2019. "The discovery of *Barretonus* (Curculionidae: Cossoninae) in the Canary Islands: barcoding, morphology and description of new species". *Acta Entomologica Musei Nationalis Pragae*, 59 (2), pp. 443-452. https://doi.org/10.2478/aemnp-2019-0033

Grancher C., Dodelin B., 2014. État des connaissances sur le genre *Cotaster* Motschulsky, 1851 en France (Coleoptera Curculionidae). *L'Entomologiste*, 70 (1), pp. 57-60.

Halmschlager E., Ladner C., Zabransky P., Schopf A., 2007. "First record of the wood boring weevil, *Pentarthrum huttoni*, in Austria (Coleoptera: Curculionidae)". *Journal of Pest Science*, 80, pp. 59-61. https://doi.org/10.1007/s10340-006-0148-3

Hammad S. M., 1955. "The immature stages of *Pentarthrum huttoni* Woll (Coleoptera : Curculionidae)". *Proceedings of the Royal Entomological Society of London (A)*, 30, pp. 33-39.

Haran J., Langor D., Roques A., Javal M., 2016. "*Pissodes irroratus* Reitter 1899, a species from East Russia new to Europe (Coleoptera: Curculionidae: Molytinae)". – CURCULIO-Institute: Germany, Mönchengladbach. *Snudebiller*, 17 (256), 6 p.

Hoffmann A., 1986. *Faune de France, 59, Coléoptères, Curculionides (Deuxième partie)*. Ed. de 1954 réimprimée, éd. Fédération française des sociétés de sciences naturelles, Paris, 1211 p.

Leather S. R., Day K. R., Salisbury A. N., 1999. "The biology and ecology of the large pine weevil, *Hylobius abietis* (Coleoptera: Curculionidae): a problem of dispersal?" *Bulletin of Entomological Research*, 89, pp. 3-16.

Nicole M.-C., Zeneli G., Lavallée R., Rioux D., Bauce É., Morency M.-J., Fenning T.-M., Séguin A., 2006. "White pine weevil (*Pissodes strobi*) biological performance is unaffected by the jasmonic acid or wound-induced defense response in Norway spruce (*Picea abies*)". *Tree Physiology*, 26, pp. 1377-1389.

Panzavolta T., Tiberi R., 2010. "Observations on the life cycle of *Pissodes castaneus* in central Italy". *Bulletin of Insectology*, 63 (1), pp. 45-50.

Sauvard D., Branco M., Lakatos F., Faccoli M., Kirkendall L., 2010. "Weevilsand Bark Beetles (Coleoptera, Curculionoidea)". Chapter 8.2., Pensoft Publishers, BioRisk, *Alien terrestrial arthropods of Europe*, 4 (1), 49 p. 978-954-642-554-6; https://doi.org/10.3897/biorisk.4.64

Tempère G., Péricart J., 1989. *Faune de France, 74, Coléoptères, Curculionidae : quatrième partie, compléments*. Ed. de 1954 réimprimée, éd. Fédération française des sociétés de sciences naturelles, Paris, 534 p.

Viiri H., Miettinen O., 2013. "Feeding Preferences of *Hylobius pinastri* Gyll". *Baltic Forestry*, 19 (1), pp. 161-164.

Yunakov N., Nazarenko V., Filimonov R., Volovnik S., 2018. "A survey of the weevils of Ukraine (Coleoptera: Curculionoidea)". *Zootaxa*, 4404 (1), pp. 1-494. https://doi.org/10.11646/zootaxa.4404.1.1

Formicidae

Wermelinger B., Düggelin C., Freitag A., Fitzpatrick B., Risch A. C., 2019. "Die Roten Waldameisen – Biologie und Verbreitung in der Schweiz". *Merkblatt für die* Praxis, 63, 12 p.

Lyctidae

Brammer A. S., 2017. "Southern *Lyctus* Beetle, *Lyctus planicollis* LeConte (Insecta: Coleoptera: Bostrichidae: *Lyctinae*)". *The Institute of Food and Agricultural Sciences (IFAS)*. EENY283, 5 p.

Gay F. J., 1953. "Observation on the biology of *Lyctus brunneus* (Steph.)". *Australian Journal of Zoology*, 1, pp. 102-140.

Henderson F. Y., 1943. "The depletion of starch from the sapwood of the ash (*Fraxinus excelsior*) and its relation to attack by powder-post beetles (*Lyctus* spp.)". *Annals of Applied Biology*, 30 (3), pp. 201-208.

Iwata R., 1986. *Mass Culture Method and Biology of the Wood-Boring Beetle, Lyctus brunneus (Stephens)*. Kyoto University, 197 p. https://doi.org/10.14989/doctor.k3625

Kartika T., Shimizu N., Yoshimura T., 2015. "Identification of Esters as Novel Aggregation Pheromone Components Produced by the Male Powder-Post Beetle, *Lyctus africanus* Lesne (Coleoptera: Lyctinae)". 13 p. PLoS ONE 10(11):e0141799. https://doi.org/10.1371/journal.pone.0141799

Kartika T., Yoshimura T., 2015. "Evaluation of Wood and Cellulosic Materials as Fillers in Artificial Diets for *Lyctus africanus* Lesne (Coleoptera: Bostrichidae)". *Insects*, 6, pp. 696-703. https://doi.org/10.3390/insects6030696

Lesne P., 1922. Régime et dégâts des Coléoptères xylophages du genre *Lyctus*. *Revue de botanique appliquée et d'agriculture coloniale*, 2e année, bulletin 12, pp. 418-420. https://doi.org/10.3406/jatba.1922.1417

May R., 2015. "*Lyctus brunneus* (Stephens, 1830), pest insects of our cultural heritage insects known to cause damage to heritage works". http://www1.montpellier.inra.fr/CBGP/insectes-du-patrimoine/?q=en/fiche-insecte/lyctus-brunneus

Parkin A., 1934. "Observation on the biology the *Lyctus* podwer beetles with special reference to oviposition and egg". *Annals of Applied Biology*, 21, pp. 495-518.

Parkin A., 1936. "A study of the food relations of the *Lyctus* powder-post beetles". *Annals of Applied Biology*, 23, pp. 369-402.

Parkin A., 1943. "The moisture content of timber in relation to attack by *Lyctus* powder-post beetles". *Annals of Applied Biology*, 30, pp. 369-400.

Rosel A., 1969. "Oviposition, egg development and other features of the biology of five species of Lyctidae (Coleoptera)". *Journal of Austr. Entomological Society*, 8, pp. 145-152.

Wilson S., 1933. "Changes in the cell contents of wood (xylem parenchyma)-relationships to the respiration of the wood resistance to *Lyctus* attack and fungal invasion". *Annals of Applied Biology*, 20, pp. 661-690.

Ptinidae

Alekseev V., Bukejs A., 2019. "*Xyletinus* (s. str.) *thienemanni* sp. nov., a new species of *Xyletininae* (Coleoptera: Ptinidae) from Eocene Baltic amber". *Acta Biol. Univ. Daugavp*, 19 (1), pp. 31-35.

Allemand R., De Laclos E., Büche B., Ponel P., 2008. Anobiidae nouveaux ou méconnus de la faune de France (3e note) (Coleoptera). *Bulletin de la Société entomologique de France*, 113 (3), pp. 397-402.

Belmain S., Blaney W. M., Simmonds M. S. J., 1998. "Host selection behaviour of deathwatch beetle, *Xestobium rufovillosum*: Oviposition preference choice assays testing old vs new oak timber, *Quercus* sp". *Entomologia Experimentalis et Applicata*, 89, pp. 193–199.

Belmain S., Simmonds M. S. J., Blaney W. M., 1999. "Deathwatch beetle, *Xestobium rufovillosum*, in historical buildings: monitoring the pest and its predators". *Entomologia Experimentalis et Applicata*, 93, pp. 97-104.

Belmain S., Simmonds M. S. J., Blaney W. M., 2002. "Influence of odor from wood-decaying fungi on host selection behavior of *Xestobium*". *Journal of Chemical Ecology*, 28 (4), pp. 741-754.

Bercedo P., Arnáiz L., 2008. "Nuevos sinónimos de *Xyletinus (Calypterus) Bucephalus Bucephalus* (Illiger, 1807) y *Calymmaderus (Calymmaderus) solidus* (Kiesenwetter, 1877) (Coleoptera: Bostrichoidea: Ptinidae: Xyletininae, Dorcatominae)". *Boletín Sociedad Entomológica Aragonesa*, 43, pp. 431-433.

Bletchly J. D., 1953. "The influence of decay in timber on susceptibility to attack by the common furniture beetle, *Anobium punctatum* de G". *Annals of Applied Biology*, pp. 218-221.

Borowski J., 2007b. "*Ptinidae (Gibbiinae and Ptininae)*". *In:* Löbl I., Smetana A. (eds.): *Catalogue of Palaearctic Coleoptera. Elateroidea - Derodontoidea - Bostrichoidea - Lymexyloidea - Cleroidea - Cucujoidea. Volume 4*. Stenstrup: Apollo Books, pp. 328-339.

Campbell W. G., Bryant S. A., 1940. "A chemical study of bearing of decay by *Phellinus cryptarum* Karst and other fungi on the destruction of wood by the death-watch beetle (*Xestobium* rufovillosum de G.)". *Biochemical Journal*, 34 (10-11), pp. 1404-1414.

Dodelin B., 2016. Sur les Episernus Paléarctiques (Col., Ptinidae, Ernobiinae). *Bulletin mensuel de la Société linnéenne de Lyon*, 85 (9-10), pp. 278-302

Español F., 1972. Note sur *Calymmaderus solidus* (Col. Anobiidae). *L'Entomologiste*, 28 (4-5), pp. 123-125.

Español F., 1977. "Los Ernobius Thoms. de la Fauna española (Col., Anobiidae, Nota 77)". *Publicaciones del Departamento de Zoologia de la Universidad de Barcelona*, 2, pp. 19-28.

Español F., 1992. *Coleoptera, Anobiidae. Fauna Iberica*. Ed. Madrid, coll. Departamento de Biologia Animal, Universitad de Barcelona, Museo Nacional de Ciencas Naturales, Consejo Superior de investigaciones cientificas, Madrid, 196 p.

Fischer R. C., 1937. "Studies of the biology *Xestobium Rufovillosum*, Summury of past work". *Annals of Applied Biology*, 24, pp. 600-613.

Fischer R. C., 1938. "Biology of the death-watch beetle, Xestobium rufovillosum. The habits of the adults". *Annals of Applied Biology*, 25, pp. 155-180.

Fischer R. C., 1941. "Studies of the biology of death-watch beetle, Xestobium rufovillosum de G. Fungal decay in timber in relation with the insect". *Annals of Applied Biology*, 27, pp. 545-557.

Fohrer F., Toriti M., Durand A., 2017. Analyse des vermoulures pour la détermination de quelques espèces d'insectes xylophages de la famille des Ptinidae. *Bulletin de la Société Entomologique de France*, 122 (2), pp. 133-142.

Gardiner P., 1953. "The Morphology and Biology of *Ernobius mollis* L. (Coleoptera-Anobiidae)". *Transactions of the Royal Entomological Society of London*, 104, pp. 1-24.

Hansen L. S., Vagn Jensen K. M., 1996. "Upper Lethal Temperature Limits of Furniture Beetle Anobium punctatum (Anobiidae)". *International Biodeterioration & Biodegradation*, 11, pp. 225-232.

Háva J., Zahradník P., 2020. "Two new species of Ptinidae (Coleoptera) from Eocene Baltic amber". *Natura Somogyiensis*, 35, pp. 5-10. https://doi.org/10.24394/NatSom.2019.35.5

Johnson C., (1975.) A review of the palaeartic species of the genus *Ernobius* Thomson (Col. Anobiidae). Entomologische Blätter, 71 (2), pp. 65-93.

Laclos E., Büche B., 2008. La vrillette sans peine : première note (coleoptera Anobiidae). *L'entomologiste*, revue d'amateurs, publiée sous l'égide de la Société entomologique de France, 1 (64), pp. 3-10.

Lohse G. A., Lucht H., 1992. "Die Käfer Mitteleuropas". *Band 13 : Supplement zu den Bänden 6-11*. Krefeld : Goecke & Evers, 176-179.

Toskina I. N., 2004. "About genera *Anobium* Fabricius, 1775, and *Cacotemnus* LeConte, 1861 (Coleoptera: Anobiidae)". *Russian Entomological Journal*, 13 (1-2), pp. 53-68.

Urban J., 2005. "Occurrence, development and harmfulness of the bark anobiid *Ernobius mollis* (L.) (Coleoptera: Anobiidae)". *Journal of Forest Science*, 51 (8), pp. 327-347.

Viñolas A., Recalde Irurzun J. I., 2018. "Los Ernobiinae de la Península Ibérica e Islas Baleares. la nota. El género *Episernus* C. G. Thomson, 1863 (Coleoptera: Ptinidae)". *Butlletí de la Institució Catalana d'Història Natural*, 82, pp. 97-107.

Viñolas A., Recalde Irurzun J. I., 2020. "Los Ptinidae (Coleoptera) de Navarra (norte de la Península Ibérica)". *Butlletí de la Institució Catalana d'Història Natural*, 84, pp. 15-24.

Zahradník P., 2007. "Contribution to knowledge of the tribe Gastrallini (Coleoptera : Bostrichoidea : Anobiidae). New species of the genus *Gastrallus* from Turkey, with review of the Palearctic species. Studies and reports of District Museum Prague-East". *Taxonomical Series*, 3 (1-2), pp. 171-178.

Zahradník P., 2013. "Two new *Ernobius* species from Cyprus (Coleoptera: Bostrichoidea: Ptinidae)". *Taxonomical Series*, 9 (2), pp. 583-590.

Kalotermitidae

Becker G., Kerner-Gang W., 1963. "Schädigung und Förderung von Termiten durch Schimmelpilze". *Journal of Applied Entomology*, (53), pp. 429-448.

Becker G., Petrowitz H. J., 1967. "Auf Termiten spurbildend wirkende Stoffe". *Naturwissenschaften*, 54, 2 p.

Becker G., 1969. "Uber einige Funde holzzerstorender Insekten auf Korsika". *Journal of Applied Entomology*, (63), pp. 93-98.

Bobadilla I., Martínez R. D., Martínez-Ramírez M., Arriaga F., 2020. "Identification of *Cryptotermes brevis* (Walker, 1853) and *Kalotermes flavicollis* (Fabricius, 1793) Termite Species by Detritus Analysis". *Forests*, 11 (408), pp. 2-10. https://doi.org/10.3390/f11040408

Buchelos C. T., Papadopoulou S., Chryssohoides C., Nota I., 2017. "List of trees and shrubs infested by *Kalotermes flavicollis* (Kalotermitidae) in Greece". *EPPO Bulletin*, 47 (2), pp. 269–273. ISSN 0250-8052. https://doi.org/10.1111/epp.12374

Cocquempot C., Valladares L., 2009. Datation des déprédations de termites et autres insectes xylophages de l'habitat et du bois d'œuvre. Approche méthodologique pour la France métropolitaine. *Cahier technique. INRA*, 67, pp. 31-42.

Evans T. A., Forschler B. T., Grace J. K., 2013. "Biology of invasive termites: a worldwide review". *Annual Review of Entomology*, 58, pp. 455-474.

Freschi M., Gigli M. C., 2008. "Un caso estremo di restauro; un San Sebastiano in legno policromo dell'ambito Silvestro dell'Aquila *Kalotermes Flavicollis*". *OPD Restauro*, (20), pp. 211-226.

Geigy R., Striebel H., 1959. "Embryonalentwicklung der Termite *Kalotermes flavicollis*". *Experentia*, 15, 2 p.

Ghesini S., Marini M., 2013. "A dark-necked drywood termite (Isoptera: Kalotermitidae) in Italy: description of *Kalotermes italicus* sp. Nov.". *The Florida Entomologist*, 96 (1), pp. 200-211.

Haverty M., Nutting W., 1974. "Natural Wood-Consumption Rates and Survival of a Dry-Wood and a Subterranean Termite at Constant Temperatures". *Annals of the Entomological Society of America*, 67, pp. 153-157.

Hickin N. E., 1971. *Termites: a world problem*. The Rentokil library, Hutchinson of London, 232 p.

Jander R., Daumer R., 1974. "Guide-line and gravity orientation of blind termites foraging in the open". *Insectes sociaux*, 21 (1), pp. 45-69.

Korb J., Hartfelder K., 2008. "Life history and development - a framework for understanding developmental plasticity in lower termites". *Biological Review* (83), pp. 295-313. https://doi.org/10.1111/j.1469-185X.2008.00044.x

Lenz M., 1972. "Über Zwischenformen bei der Kastenbildung von *Kalotermes flavicollis* (F.) und *Heterotermes indicola* (Wasm.) (Kalotermitidae, Rhinotermitidae)". *Journal of Applied Entomology*, (72), pp. 113-115.

Luchetti A., 2020. *Drywood Termites (Kalotermitidae). In : Encyclopedia of Social Insects*, Springer, 3 p.

Pearce M. J., 2000. *Termites: biology and pest management*. Formerly of the Natural Resources institute Chatham, Kent, UK. CAB International, Cambridge, 142 p.

Pervez A., 2018. *"Termite Biology and Social Behaviour"*. *In*: Khan A., Ahmad W., (eds) *Termites and Sustainable Management*. Springer, pp. 119-143.

Springhetti A., Sita E., 1989. "Influence of reproductives on tunnelling behaviour in *Kalotermes flavicollis* Fabr.". *Insectes Sociaux* 36 (1), pp. 70-73.

Su N. Y., Scheffrahn R., 2001. *"Termites as Pests of Buildings"*. *In*: Abe T., Bignell D. E., Higash M., *Termites Evolution, Sociality, Symbioses, Ecology*. Springer, pp. 437-453.

Scolytinae

Anderbrant O., 1985. "Dispersal of reemerged spruce bark beetles, *Ips typographus* (Coleoptera, Scolytidae): a mark-recapture experiment". *Zeitschrift für Angewandte Entomology*, 99, pp. 21-25.

Annila E., 1969. "Influence of temperature upon the development and voltinism of *Ips typographus* L. (Coleoptera, Scolytidae)". *Annales Zoologici Fennici*, 6, pp. 161-208.

Balachowsky A., 1997. *Faune de France, 50, Coléoptères scolytides*. Ed. de 1949 réimprimée, éd. Fédération française des sociétés de sciences naturelles, Office central de faunistique, dir. Honoraire : P. de Beauchamp, L. Chopard, librairie de la Faculté des Sciences, CNRS, Paris, 320 p.

Chararas C., 1962. *Scolytides des conifères*. Lechevalier, Paris, 556 p.

Christiansen E., Bakke A., 1988. *"The spruce bark beetle of Eurasia"*. *In*: Berryman A. A. (ed) *Dynamics of forest insects populations*. Plenum Press, New York, pp. 479-503.

De Jong M. C. M., Grijpma P., 1986. "Competition between larvae of *Ips typographus*". *Entomologia Experimentalis et Applicata*, 41, pp. 121-133.

Forster B., Meier F., 2010. "Sturm, Witterung und Borkenkäfer. Risikomanagement im Forstschutz". *2. Aufl. Merkbl. Prax.*, 44, 8 p.

Forster B., 2017. "Kupferstecher und Furchenflügeliger Fichtenborkenkäfer". *Merkbl. Prax.* 58, 8 p.

Galko J., Pavlík Š., 2009. "Parasitic wasps (Hymenoptera) of the European oak bark beetle larvae (*Scolytus intricatus* Ratz., Coleoptera: Scolytidae). Lesn. Čas." *Forestry Journal*, 55 (1), pp. 1-12.

Gurevitz E., Ledoux A., 1981. Attraction exercée par les plantes-hôtes sur le scolyte méditerranéen, *Scolytus* (*Ruguloscolytus*) *mediterraneus* Eggers. *Agronomie, EDP Sciences*, 1 (3), pp. 249-254. hal-00884251

Holuša J., Lukášová K., Grodzki W., Kula E., Matoušek P., 2012. "Is *Ips amitinus* (Coleoptera: Curculionidae) Abundant in Wide Range of Altitudes?" *Acta zool. bulg.*, 64 (3), pp. 219-228.

Jankowiak R., 2005. "Fungi associated with *Ips typographus* on *Picea abies* in southern Poland and their succession into the phloem and sapwood of beetle-infested trees and logs". *Forest Pathology*, 35, pp. 37-55.

Kula E., Ząbecki W., 2006. "Spruce windfalls and cambioxylophagous fauna in an area with the basic and outbreak state of *Ips typographus* (L.)". *Journal of Forest Science*, 52 (11), pp. 497-509.

Kula E., Ząbecki W., 2010. "Merocoenoses of cambioxylophagous insect fauna of Norway spruce (*Picea abies* [L.] Karst.) with focus on bark beetles (Coleoptera: Scolytidae) and types of tree damage in different gradation conditions". *Journal of Forest Science*, 56 (10), pp. 474-484.

Lieutier F., Levieux J., 1985. Les relations conifères-scolytides : importance et perspectives de recherches. *Annals of Forest Research*, 42 (4), pp. 359-370.

Lieutier F., Yart A., Salle A., 2009. "Stimulation of tree defenses by Ophiostomatoid fungi can explain attack success of bark beetles on conifers". *Annals of Forest Science*, 66 (801), pp. 2-22. https://doi.org/10.1051/forest/2009066

Lukášová K., Holuša J., Turčáni M., 2013. "Pathogens of *Ips amitinus*: new species and comparison with *Ips typographus*". *J. Appl. Entomol.*, 137, pp. 188-196.

Marković Č., Stojanović A., 2012. "Fauna of phloemo-xylophagous insects, their parasitoids and predators on *Ulmus minor* in Serbia". *Biologia, Section Zoology*, 67 (3), pp. 584-589. https://doi.org/10.2478/s11756-012-0044-7

Menkis A., Östbrant I. L., Davydenko K., Bakys R., Balalaikins M., Vasaitis R., 2016. "*Scolytus multistriatus* associated with Dutch elm disease on the island of Gotland: phenology and communities of vectored fungi". *Mycol Progress*, 15, 55 p. https://doi.org/10.1007/s11557-016-1199-3

Nierhaus-Wunderwald D., 1995. "Der Grosse Lärchenborkenkäfer. Biologie, Überwachung und forstliche Massnahmen". *Merkbl. Prax.* 24, 6 p.

Nierhaus-Wunderwald D., 1996. "Die natürlichen Gegenspieler der Borkenkäfer". *2. Auflage. Merkbl. Prax.* 19, 8 p.

Nierhaus-Wunderwald D., Forster B., 2000. "Rindenbrütende Käfer an Föhren". *2. Auflage. Merkbl. Prax.* 31, 12 p.

Nierhaus-Wunderwald D., 2004. "Biologie der Buchdruckerarten". *3. überarbeitete Aufl. Merkbl. Prax.* 8 p.

Persson Y., Vasaitis R., Långström B., Öhrn P., Ihrmark K., Stenlid J., 2009. "Fungi Vectored by the Bark Beetle *Ips typographus* following Hibernation Under the Bark of Standing Trees and in the Forest Litter". *Microb. Ecol.*, 58, pp. 651-659. https://doi.org/10.1007/s00248-009-9520-1

Piou D., Lieutier F., Yart A., 1989. Observations symptomatologiques et rôles possibles d'*Ophiostoma minus* Hedgc. (ascomycète : Ophiostomatales) et de *Tomicus piniperda* L. (Coleoptera : Scolytidae) dans le dépérissement du pin sylvestre en forêt d'Orléans. *Annales des sciences forestières*, INRA/EDP Sciences, 46 (1), pp. 39-53. hal-00882463

Polyanina K. S., Mandelshtam M. Y., Ryss A. Y., 2019. "Brief Review of the Associations of Xylobiont Nematodes with Bark Beetles (Coleoptera, Curculionidae: Scolytinae)". *Entomological Review*, 99 (5), pp. 598-614.

Procházka J., Schlaghamerský J., 2019. "Does dead wood volume affect saproxylic beetles in montane beech-fir forests of Central Europe?" *Journal of Insect Conservation*, 23, pp. 157-173. https://doi.org/10.1007/s10841-019-00130-4

Rigling D., Hilfiker S., Schöbel C., Meier F., Engesser R., Scheidegger C., Stofer S., Senn-Irlet B., Queloz V., 2016. "Das Eschentriebsterben. Biologie, Krankheitssymptome und Handlungsempfehlungen". *Merkbl. Prax.* 57, 8 p.

Sallé A., Baylac M., Lieutier F., 2005. "Size and shape changes of *Ips typographus* L. (Coleoptera: Scolytinae) in relation to population level". *Agricultural and Forest Entomology*, 7, pp. 297-306.

Sanchez A., Chittaro Y., Germann C., Knížek M., 2020. "Annotated checklist of *Scolytinae* and *Platypodinae* (Coleoptera, Curculionidae) of Switzerland". *Alpine Entomology*, 4, pp. 81-97. https://doi.org/10.3897/alpento.4.50440

Schimitschek E., 1931. "Von Der achtzähnige Lärchenborkenkäfer *Ips cembrae* Heer". *Journal of Applied Entomology*, 17, pp. 253-344.

Stauffer C., Lakatos F., Hewitt G. M., 1999. "Phylogeography and postglacial colonization routes of *Ips typographus* L. (Coleoptera, Scolytidae)". *Molecular Ecology*, 8, pp. 763-773.

Takov D., Doychev D., Linde A., Draganova S., Pilarska D., 2011. "Pathogens of bark beetles (Coleoptera: Curculionidae) in Bulgarian forests". *Phytoparasitica*, 39, pp. 343-352. https://doi.org/10.1007/s12600-011-0167-3

Vasiliauskas R., Menkis A., Finlay R. D., Stenlid J., 2007. "Wood-decay fungi in fine living roots of conifer seedlings". *New Phytologist*, 174, pp. 441-446.

Warzée N., Grégoire J.-C. 2013. "Biodiversité forestière et ennemis naturels des scolytes : le cas exemplaire de *Thanasimus formicarius*". *Forêt Wallonne*, 66, pp. 2-6.

Wermelinger B., 2004. "Ecology and management of the spruce bark beetle *Ips typographus* - a review of recent research". *Forest Ecology and Management*, 202, pp. 67-82.

Weslien J., 1992. "Monitoring *Ips typographus* (L.) populations and forecasting damage". *Journal of Applied Entomology*, 114, pp. 338-340.

Zumr V., 1992. "Dispersal of the spruce bark beetle *Ips typographus* (L.) (Col., Scolytidae) in spruce woods". *Journal of Applied Entomology*, 114, pp. 348-352.

Web Resources

CABI - https://www.cabi.org/isc/
Cerambycidae - http://www.cerambyx.uochb.cz/
CICRP - http://insectes-nuisibles.cicrp.fr/
Ephytia - http://ephytia.inra.fr/
Fauna Europea - https://fauna-eu.org/
Gbif - https://www.gbif.org/fr/
INPN MNHM - https://inpn.mnhn.fr/accueil/index
Itis - https://www.itis.gov/
Mymecofourmis - https://www.myrmecofourmis.com/fiches/673366#biologie-ecologie
UKbeetles - https://www.ukbeetles.co.uk/

Lightning Source UK Ltd.
Milton Keynes UK
UKHW050623250722
406326UK00001B/2

9 783030 66391